MINING FUNDAMENTALS

BASIC CYANIDE CHEMISTRY

An Essential Guide for Gold and Silver Leaching

MICHAEL M. BOTZ

PUBLISHED BY THE SOCIETY FOR MINING, METALLURGY & EXPLORATION

Society for Mining, Metallurgy & Exploration (SME)
12999 E. Adam Aircraft Circle
Englewood, Colorado, USA 80112
(303) 948-4200 / (800) 763-3132
www.smenet.org

SME advances the worldwide mining and underground construction community through information exchange and professional development.

ISBN 978-0-87335-523-0
Ebook 978-0-87335-526-1

Library of Congress Cataloging-in-Publication Data has been applied for

13 Sulfur Reactions 67
14 Cyanide Analysis 73
Classifications of Cyanide 73
WAD and Total Cyanide 73
Silver Nitrate Titration 75
15 Properties of Sodium Cyanide 81
16 Cyanide Safety 87
Supplemental Data 89
Further Reading 93
References 93
Index 99

Contents

Acknowledgments v

1 Introduction 1
- Tribute to Cyanide 1
- Cyanide Defined 2
- Industrial Forms of Cyanide 2
- Metal-Cyanide Complexes 3
- Cyanate and Thiocyanate 4
- Other Metal-Ligand Complexes 5
- Stability Constants 5
- Eh-pH Diagrams 5

2 Free Cyanide 9
- Free Cyanide Speciation 9
- Hydrogen Cyanide Volatilization 10
- Simple Cyanides 12
- Unit Conversions 12

3 Gold 15

4 Silver 19

5 Iron 23

6 Cobalt 29

7 Nickel 33

8 Copper 37

9 Zinc 47

10 Cadmium 55

11 Mercury 59

12 Mixed Complexes 63

Acknowledgments

The author is grateful for the helpful insights, reviews, and data provided by the following friends and colleagues: John Marsden, Paul Breuer, William Lipps, Rob Dunne, Matthew Jeffrey, Roy Norcross, John Wilmot, and Anthony Staley. Your time, generosity, and expertise are valued.

1 Introduction

This handbook provides an introduction to the chemistry of cyanide as it relates to gold and silver leaching. It is intended as a reference for those involved with designing, operating, and troubleshooting the processes used for leaching and recovering gold and silver. Basic knowledge about the chemistry of cyanide enhances our understanding of these processes and how they are affected by the chemical characteristics of leach solutions and the ores being processed. The subject matter is a convenient compilation of practical information about cyanide and its interactions with gold, silver, and several other metals. Readers are presumed to have a background in general chemistry and basic familiarity with the common processes used for leaching and recovering gold and silver.

The work opens with some background material and descriptions of the chemistry of cyanide in the absence of metals (Chapters 1–2). It then continues with a discussion about the interaction of cyanide with gold and silver (Chapters 3–4) and several other metals often encountered when gold and silver are being leached (Chapters 5–11). A detailed examination of a particular leach solution follows, taking into consideration a comprehensive solution analysis (Chapter 12). The reactions of sulfur species are then considered with respect to the high consumptions of cyanide and oxygen that are sometimes observed when leaching ores containing sulfide minerals (Chapter 13). General classifications of cyanide species and commentary about their analysis are then presented (Chapter 14). The work concludes with a compilation of the physical properties of sodium cyanide and some general aspects of cyanide safety (Chapters 15–16). Although this handbook is intended as an introductory text, many technical references are cited should the reader wish to pursue a more in-depth study of a particular topic (see "Further Reading" and "References" at the end of the book).

TRIBUTE TO CYANIDE

Cyanide has been used worldwide for gold and silver leaching for well over 125 years. Its effectiveness and efficiency at leaching gold and silver is quite noteworthy. Ores containing only a few parts per million of gold or silver, and in some cases, fractions of a part per million of gold, can routinely be leached with recoveries approaching 75%–90% or higher. The durations required for leaching are measured in hours for agitated tank operations and in weeks to months for heap leach operations. All of this occurs with solutions containing only a fractional percent of cyanide, solutions that are safely and responsibly managed at hundreds of operations worldwide. Many innovative and effective leaching and recovery processes have been developed, including those for the downstream detoxification and recycling of cyanide. Doubtless more will be developed in the future. The chemistry, toxicity, environmental behavior, and analysis of cyanide are all well understood. Many professionals in mining companies, engineering firms, metallurgical testing laboratories, universities, and consultancies regularly share their experiences at conferences and through technical publications. Add to this that cyanide is widely available and relatively inexpensive, and the result is an impressive chemical that is responsible for a majority share of gold and silver production worldwide. Many millions of ounces of gold and silver have been produced through cyanide leaching, and the world can look forward to many millions more in the coming years.

CYANIDE DEFINED

As a starting point, the term "cyanide" should be defined. Cyanide consists of one atom of nitrogen triple-bonded to one atom of carbon, the resulting molecule having a single negative charge. The chemical structure used to represent the cyanide molecule is shown in Figure 1.1.

$$^{-}C\equiv N$$

FIGURE 1.1 Structure of the cyanide molecule

The word cyanide originates from the Greek κύαηος, *kyanos*, meaning "dark blue." The long-known cyanide compounds of Prussian blue and Turnbull's blue have intense dark blue colors, and this property has made them useful as dyes and pigments.

Cyanide falls into a class of chemical compounds known as *ligands*. A ligand forms chemical bonds with metals, and in many cases, the resulting metal-ligand complex is soluble in water. A fitting example is gold, which reacts with cyanide to form the soluble gold-cyanide complex $Au(CN)_2^-$. In the absence of a ligand such as cyanide, gold remains insoluble in water, and leaching does not occur. The behavior of cyanide with silver is similar. The ability of cyanide to leach gold and silver effectively and efficiently has led to its widespread use and importance in the precious metals mining industry.

INDUSTRIAL FORMS OF CYANIDE

At gold and silver leaching operations, cyanide is almost exclusively supplied in the form of sodium cyanide (NaCN). This is due to the economics associated with sodium cyanide production. The first step in the production process involves the generation of hydrogen cyanide (HCN) gas, which is then absorbed into a concentrated solution of sodium hydroxide (NaOH). The product is a concentrated solution of sodium cyanide that includes a small residual of sodium hydroxide. Sodium hydroxide is used for the absorption step because of its availability, ease of handling, high solubility in water, and relatively low cost.

Sodium cyanide can be sold as a concentrated liquid or can be crystallized and then sold in solid form. Potassium cyanide (KCN), calcium cyanide ($Ca(CN)_2$), and several other cyanide salts are also available but these are mostly used in laboratory or specialty chemistry applications. Many metal-cyanide solids are available, such as copper cyanide (CuCN) and potassium hexacyanoferrate (III) ($K_3Fe(CN)_6$). These metal-cyanide compounds are used primarily in laboratory or specialty chemistry applications as well. Several useful physical properties of sodium cyanide are provided in Chapter 15.

METAL-CYANIDE COMPLEXES

A wide variety of cyanide complexes can be present in leach solutions and a partial list is provided in Table 1.1. The presence of any of these complexes in a leach solution depends on the occurrences of various metals and minerals in the material being leached and their inherent reactivities toward cyanide. The list of metal-cyanide complexes in Table 1.1 is not exhaustive but includes those most often encountered in gold and silver leaching. Many other complexes are possible, including cyanide bound with more than one metal and metals bound to other ligands. However, in most cyanide leach circuits these other complexes are usually not present at elevated concentrations and are not therefore of primary importance.

TABLE 1.1 Some common aqueous metal-cyanide complexes

Metal	Complexes	Discussion
None	CN^- $HCN_{(aq)}$	Chapter 2
Gold (Au^+)	$Au(CN)_2^-$	Chapter 3
Silver (Ag^+)	$Ag(CN)_2^-$ $Ag(CN)_3^{2-}$	Chapter 4
Iron (Fe^{2+}) Iron (Fe^{3+})	$Fe(CN)_6^{4-}$ $Fe(CN)_6^{3-}$	Chapter 5
Cobalt (Co^{3+})	$Co(CN)_6^{3-}$	Chapter 6
Nickel (Ni^{2+})	$NiH_2(CN)_4^0$ $NiH(CN)_4^-$ $Ni(CN)_4^{2-}$	Chapter 7
Copper (Cu^+)	$Cu(CN)_2^-$ $Cu(CN)_3^{2-}$ $Cu(CN)_4^{3-}$	Chapter 8
Zinc (Zn^{2+})	$ZnCN^+$ $Zn(CN)_2^0$ $Zn(CN)_3^-$ $Zn(CN)_4^{2-}$	Chapter 9
Cadmium (Cd^{2+})	$CdCN^+$ $Cd(CN)_2^0$ $Cd(CN)_3^-$ $Cd(CN)_4^{2-}$	Chapter 10
Mercury (Hg^{2+})	$HgCN^+$ $Hg(CN)_2^0$ $Hg(CN)_3^-$ $Hg(CN)_4^{2-}$	Chapter 11

A Sharp Chemist

The chemistry of cyanide can become quite complicated as chemists investigate the very detailed aspects of its behavior. The brilliant chemist A.G. Sharpe (1976) wrote about some aspects of the chemistry of cyanide complexes:

> Cyanide ion is isoelectronic with carbon monoxide, molecular nitrogen, and the C_2^{2-} and NO^+ ions. The electronic configuration for these species is $(\sigma 1s)^2(\sigma^* 1s)^2(\sigma 2s)^2(\sigma^* 2s)^2(\sigma 2p_x)^2(\pi 2p_y)^2(\pi 2p_z)^2$ and the two atoms are triply bonded (one σ- and two π-bonds). Cyanide ion acting as a monodentate ligand always appears to have carbon as the donor atom. Cyanide can, however, also act as an ambidentate ligand, i.e., act as a donor at both ends of the ion. This happens not only in polymeric species such as silver(I) cyanide, gold(I) cyanide, and complexes of the Prussian blue type, but also in boron trifluoride adducts such as $K_2[Ni(CN)_4]{\cdot}4BF_3$ and $K_4[Fe(CN)_6]{\cdot}6BF_3$. It is generally accepted that, like carbon monoxide in the metal carbonyls, cyanide in complex cyanides acts as a σ-donor and a π-receptor; in octahedral complexes the d_{xy}, d_{xz}, and d_{yz} orbitals, are suitably orientated for overlap with a π*-antibonding orbital of the ligand.

Nothing nearly as complicated as this will be discussed in this handbook.

Typically, the concentrations of gold, silver, cobalt, nickel, cadmium, and mercury are relatively low in leach solutions, usually less than about 5–20 mg/L and in many cases <1 mg/L. It is not unusual for leach solutions to contain moderate to relatively high concentrations of copper or zinc; in some cases, many hundreds or even thousands of milligrams per liter may be present. Iron tends to be present at intermediate to relatively low concentrations, often in the range of about 1–50 mg/L. Ores with sulfidic minerals and flotation concentrates are often associated with higher levels of metals in solution, as compared to ores that are oxide in nature.

CYANATE AND THIOCYANATE

In cyanide leach circuits, there may be some amounts of thiocyanate (SCN^-) or cyanate (OCN^-) present in solution. These compounds both contain C and N in their structures, which can give the impression they are cyanide compounds. But in fact, thiocyanate and cyanate are not cyanide compounds and their chemistry, analysis, and toxicity are quite different from cyanide. The chemical structures used to represent thiocyanate and cyanate are shown in Figure 1.2. Both of these molecules have a single negative charge and are highly soluble in water. Selenocyanate ($SeCN^-$) is a similar compound that can form through the reaction of cyanide with selenium-bearing minerals.

The reaction of cyanide with sulfide minerals is common and thiocyanate can form as a result. This topic is addressed in Chapter 13. In the absence of a strong oxidant, cyanide is only slowly oxidized to cyanate. But the reaction is catalyzed by galvanic surfaces, such as

$$^{-}S-C\equiv N$$
$$^{-}O-C\equiv N$$

FIGURE 1.2 Structures of the thiocyanate and cyanate molecules

sulfide minerals and activated carbon, the rate of which is increased by the presence of copper. In cyanide detoxification circuits, cyanide is intentionally oxidized to cyanate using stronger oxidants and often with a catalyst. With respect to gold and silver leaching chemistry, the presence of thiocyanate or cyanate at moderate levels is not detrimental. But their presence at elevated levels, particularly thiocyanate, can be indicative of side reactions that are occurring to consume cyanide.

OTHER METAL-LIGAND COMPLEXES

The metals discussed in this handbook can form soluble complexes with other ligands, such as hydroxide, thiosulfate, thiocyanate, ammonia, chloride, sulfide, and sulfate. If there are elevated levels of these other ligands in solution, they should be considered along with cyanide. However, in most leach solutions, the metal-cyanide complexes discussed in this handbook predominate.

STABILITY CONSTANTS

Throughout this handbook, there is reference to the term *stability constant*, denoted by the symbol β. In short, the stability constant is a measure of the extent to which a metal-cyanide complex dissociates into its individual constituents. The following is an example for the copper-cyanide complex $Cu(CN)_2^-$, which is composed of Cu^+ and CN^-.

$$Cu^+ + 2CN^- \rightleftharpoons Cu(CN)_2^-$$

The stability constant for this particular equilibrium reaction is 8.71×10^{23} at 25°C.* This value can be used to calculate the individual concentrations of Cu^+, CN^-, and $Cu(CN)_2^-$ for a particular solution. A convenient way of expressing the values of stability constants is by using log β, which for this case the value is 23.94. It is beyond the scope of this handbook to go into further detail, but Baes and Mesmer (1976) and Fuerstenau and Han (2003) and other basic chemistry texts provide good coverage of the concept of stability constants.

Eh-pH DIAGRAMS

It is helpful to utilize Eh-pH diagrams when discussing the various complexes that can be present in cyanide leach systems.† The concept of pH is generally understood, but the concept of Eh is less so. Eh is a measure of the potential for a chemical system to undergo either oxidation

* Ranges of reported values exist for all stability constants, so the value presented here should not be interpreted as being the only correct value. For this particular reaction, values of log β ranging from 16.3 to 24.9 were located in published literature. But the more reliable values appear to be 23 ± 1.5. Values for log β selected by the author to perform equilibrium calculations for this handbook are listed in each chapter along with references to the original sources of data.

† Eh-pH diagrams are commonly referred to as *Pourbaix diagrams*, in namesake of the chemist Marcel Pourbaix and his well-known text *Atlas of Electrochemical Equilibria in Aqueous Solutions* (Pourbaix 1974).

reactions or reduction reactions (hence the origin of the term *redox* to describe a class of chemical reactions). Familiar oxidants include oxygen, chlorine, ozone, and hydrogen peroxide, and these substances can promote the oxidation of compounds. Reduction is simply the opposite of oxidation. Familiar reducing agents include sulfide, sulfite, sulfur dioxide, and elemental zinc metal. Eh has the unit of volts, and higher values indicate more oxidizing conditions while lower values indicate more reducing conditions. This is important at gold and silver leaching operations where both oxidation and reduction reactions take place.

There are some aspects of Eh-pH diagrams that are helpful to understand. The following discussion in three points relates to the Eh-pH diagrams for the cyanide-water system presented in Figure 1.3.

1. Each designated region on an Eh-pH diagram identifies the single *predominant* complex that exists within the indicated boundaries. In Figure 1.3(a), to the left of pH 9.2 hydrogen cyanide ($HCN_{(aq)}$) is the predominant complex in solution, while to the right of pH 9.2, the cyanide anion (CN^-) is the predominant complex in solution. However, this does not imply that *only* these complexes are present in these regions; the diagram simply shows the *predominant* complex. For example, at pH 10, free cyanide in solution consists of approximately 15% as $HCN_{(aq)}$ and 85% as CN^-. The predominant complex at pH 10 is therefore CN^-, but at the same time, an important amount of $HCN_{(aq)}$ is also present. For this reason, pH-concentration diagrams are used in this handbook, along with Eh-pH diagrams, to show the relative proportions of various complexes in solution under given Eh-pH conditions.
2. Eh-pH diagrams show the complexes that exist at *equilibrium* after all the prescribed chemical reactions have come to completion. Figure 1.3(b) shows the complexes present in the cyanide-water system if cyanate (OCN^-) is included in the calculations. In this case, it is shown thermodynamically that cyanate has a large region of stability as compared to cyanide and that it predominates when equilibrium is achieved. Cyanide is shown to be thermodynamically stable only under strongly reducing conditions. Fortunately, the *kinetics* for the conversion (oxidation) of cyanide to cyanate are slow in most situations, and this is what enables the use of cyanide for gold and silver leaching. Thus, Figure 1.3(a) is most relevant when considering free cyanide in leaching systems. For practical purposes, cyanate formation does not occur except in the presence of a strong oxidant, a catalyst, long reaction times, or elevated temperatures. And actually, if cyanide were allowed to come to its final state of equilibrium, considering the most stable of all the possible thermodynamic species, the carbon portion of cyanide would convert to carbon dioxide (CO_2) or methane (CH_4), while the nitrogen portion would convert to molecular nitrogen (N_2), ammonia (NH_3), or nitrate (NO_3^-). The Eh-pH diagrams presented in the following chapters of this handbook exclude cyanate as a possible compound.
3. As a final note, there are upper and lower dashed lines typically shown on Eh-pH diagrams. These lines indicate the boundary limits for oxidation and reduction of water. The upper line shows oxidizing conditions where water breaks down to form molecular oxygen (O_2), while the lower line shows the reducing conditions where water breaks down to form molecular hydrogen (H_2). For cyanide leaching systems, the region for chemistry essentially exists between the two stability lines. However, the chemistry above

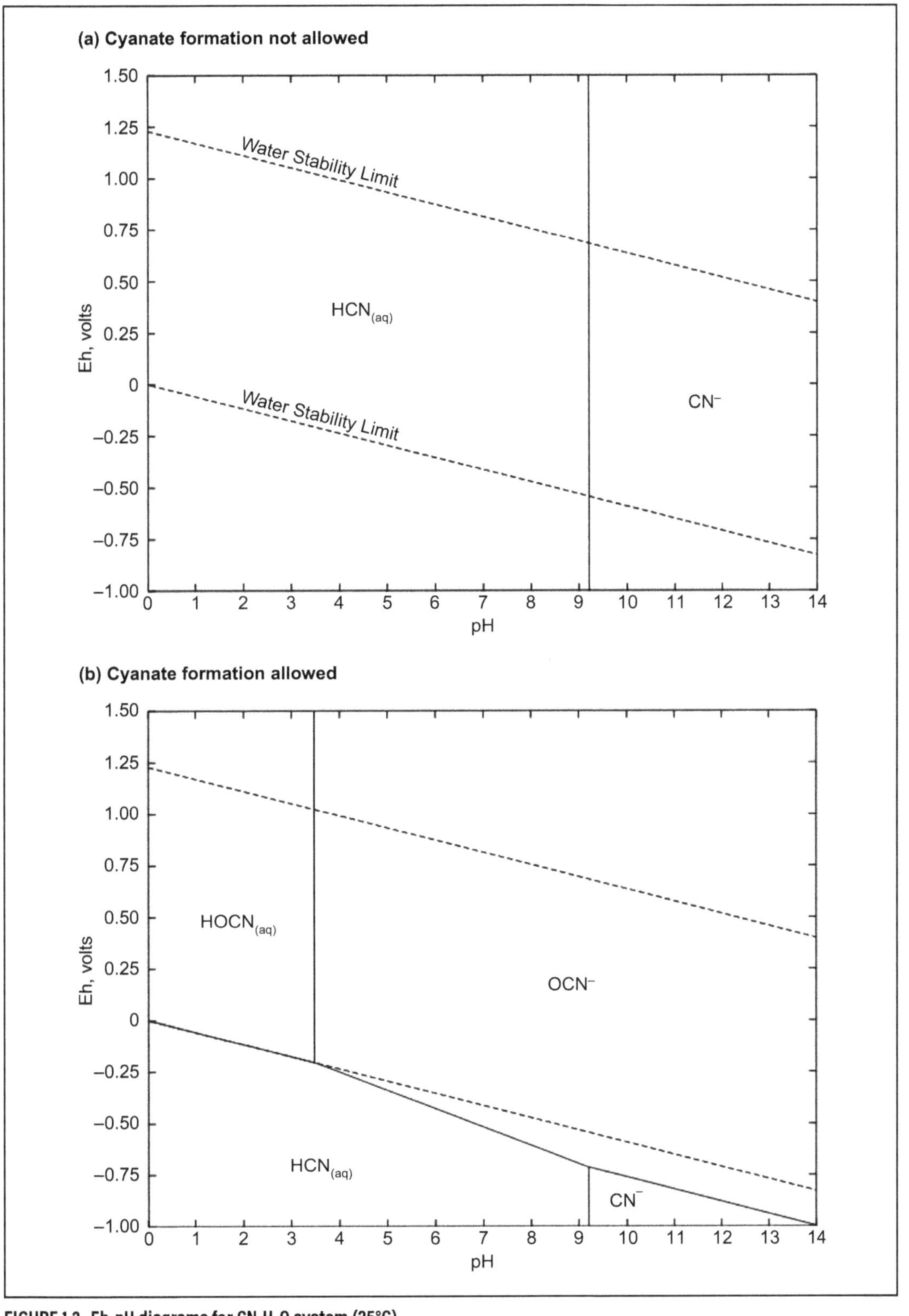

FIGURE 1.3 Eh-pH diagrams for CN-H_2O system (25°C)

Cyanide Mountain

Cyanide Mountain is located in the rugged terrain of western Montana (United States) at the edge of the Scapegoat Wilderness area. The source of the interesting name for this mountain is not known. The elevation of the peak is 2,408 m (7,900 ft), and it is a strenuous climb. The author and his wife made a partial ascent in 2023 but were repelled by the steepness and heavy timber downfall. The next day, a long hike to Elbow Creek several kilometers to the west was successful. The author's consulting practice is named Elbow Creek Engineering, hence the interest in the coincidental names of these two landmarks.

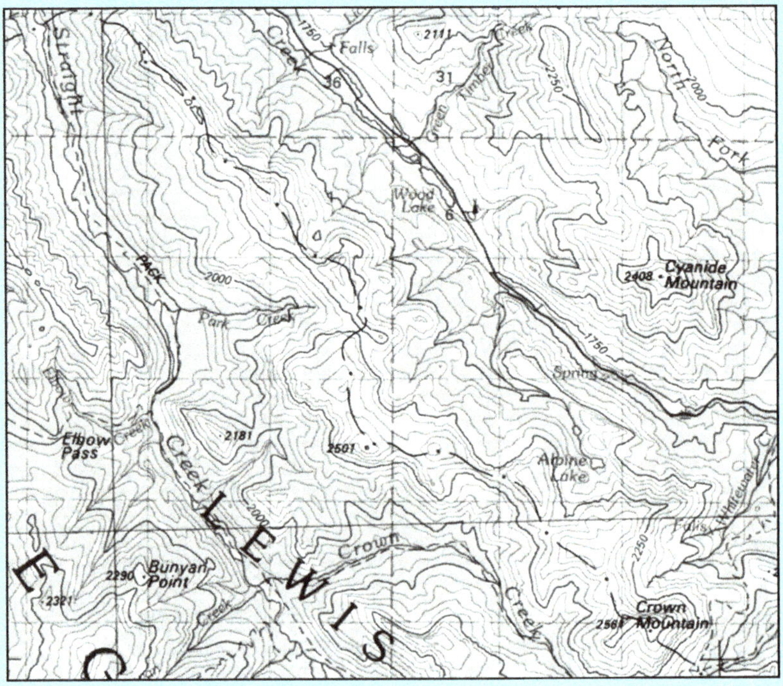

Courtesy of U.S. Geological Survey

Portion of a 1993 topographic map for Dearborn River, Montana, United States

and below the stability lines is interesting and often aids in understanding how systems may respond under strongly oxidizing or reducing conditions.

The Eh-pH diagrams presented in this handbook were generated using the Geochemist's Workbench (Aqueous Solutions 2011–2023) with a custom thermodynamic database compiled by the author.

6	7
C	N
Carbon	Nitrogen
12.0107	14.0067

Free Cyanide

2

FREE CYANIDE SPECIATION

Free cyanide, by definition, includes only the cyanide anion CN^- and aqueous hydrogen cyanide $HCN_{(aq)}$. Any other cyanide complex that may be present in solution is not considered free cyanide. The relative proportions of CN^- and $HCN_{(aq)}$ in solution are primarily a function of pH and secondarily a function of temperature. Figure 2.1 shows the proportions according to a range of pH values and temperatures.

Most cyanide leach circuits are operated around pH 10.5 ± 0.5, although lower pH values may be encountered. For example, in Western Australia, the brackish water supplies often have a high magnesium content, which limits the practical cyanidation pH to about 9.2–9.5. Magnesium hydroxide begins to precipitate above pH 9.5, thereby consuming significant quantities of lime. At a pH of 10.5, approximately 95% of cyanide is present as CN^- and 5% as $HCN_{(aq)}$ (25°C). This pH level provides for relatively low concentrations of $HCN_{(aq)}$ and limits the amount of cyanide that may be lost through volatilization. Importantly, the compound $HCN_{(aq)}$ refers to aqueous hydrogen cyanide (i.e., hydrogen cyanide that is dissolved in solution). Hydrogen cyanide is quite soluble in water and only slowly volatilizes to gaseous form unless the solution is agitated or aerated.

The equilibrium reaction for free cyanide in solution is written as follows:

$$H^+ + CN^- \rightleftharpoons HCN_{(aq)}$$

Le Chatelier's principle suggests at acidic pH values (i.e., high concentrations of H^+ in solution), the preceding reaction shifts to the right to form $HCN_{(aq)}$. And, of course, that is what occurs in practice and one of the reasons cyanide leach circuits are operated under alkaline pH conditions (i.e., with lower concentrations of H^+ in solution).

Stability constants for the preceding reaction are shown in Table 2.1 at various temperatures. Note that lower solution temperatures shift the equilibrium to slightly favor the formation of $HCN_{(aq)}$ at a given pH. Higher levels of salinity act to lower the value of the stability constant.

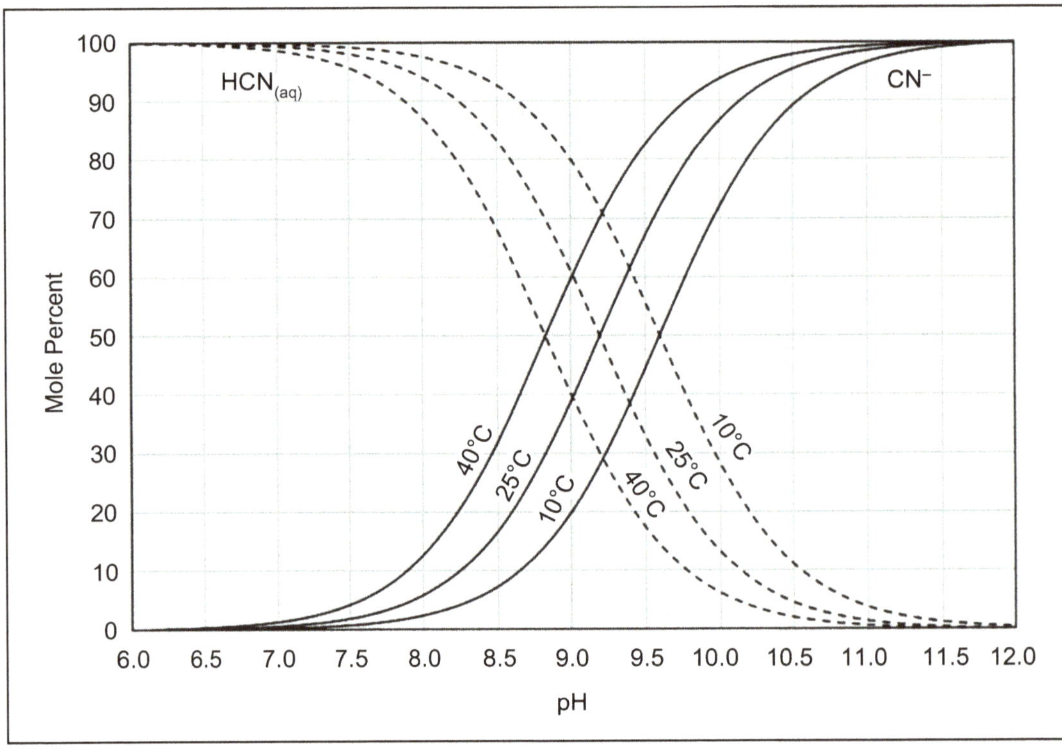

FIGURE 2.1 Distribution of free cyanide according to pH and temperature

TABLE 2.1 Stability constants for free cyanide*

Equilibrium Reaction $H^+ + CN^- \rightleftharpoons HCN_{(aq)}$	
Temperature, °C	**log β**
0	10.00
10	9.66
25	9.22
50	8.63
100	7.92

* Calculated using the relationship in Edwards et al. 1978.

HYDROGEN CYANIDE VOLATILIZATION

Cyanide that is present in solution as $HCN_{(aq)}$ can volatilize to yield the gas $HCN_{(g)}$. The rate of volatilization depends primarily on the solution temperature and the degree of agitation and aeration. The equilibrium reaction for volatilization is written as follows:

$$HCN_{(aq)} \rightleftharpoons HCN_{(g)}$$

This reaction is rarely at equilibrium in an open cyanide leach system because the rate of gas dispersion into the surrounding air is generally much faster than the rate of volatilization. But in practice, some amount of hydrogen cyanide does volatilize from leach solution. The

Cyanide for Plastics Production

Hydrogen cyanide production is a precursor to the manufacture of sodium cyanide that is used for gold and silver leaching. But the manufacture of sodium cyanide consumes only about 8% of the global production of hydrogen cyanide. Other major users of hydrogen cyanide include the manufacture of adiponitrile (47%), acetone cyanohydrin (27%), and methionine (6%). Manufacturing these products, plus sodium cyanide manufacturing, consumes about 88% of the total produced hydrogen cyanide. Adiponitrile is a key ingredient in the production of nylon. Acetone cyanohydrin is used to make methyl methacrylate, which is then used to make acrylic sheeting. Methionine is an essential amino acid used in animal feed. There are currently six different industrial processes utilized to produce a majority of hydrogen cyanide, most of which is used directly on site in the plastics industry.

Source: Maxwell 2004

$HCN_{(g)}$ concentrations present in working areas are normally low, provided the leach solution pH is sufficiently alkaline and work areas are open or suitably ventilated. The Eh-pH diagram for free cyanide is discussed in Chapter 1 (see the "Eh-pH Diagrams" section).

At equilibrium, the concentration of $HCN_{(g)}$ above a solution containing cyanide is directly proportional to the concentration of $HCN_{(aq)}$ in solution, which in turn is a function of pH. Henry's law is used for relatively dilute solutions:

$$p[\mathrm{HCN}_{(g)}] = K_{\mathrm{H}} \times [\mathrm{HCN}_{(aq)}]$$

With the preceding relationship, $p[HCN_{(g)}]$ is the partial pressure of $HCN_{(g)}$ above the solution, K_H is the Henry's law constant, and $[HCN_{(aq)}]$ is the concentration of $HCN_{(aq)}$ in solution. The Henry's law constant is quite sensitive to temperature, as shown in Table 2.2. Higher temperatures correspond to an increased tendency for $HCN_{(aq)}$ to volatilize from solution because of the higher vapor pressure.

TABLE 2.2 Henry's law constants as a function of temperature*

	Henry's Law Constant	
Temperature, °C	kg-atm/mole	atm/mole fraction
0	0.015	0.83
10	0.033	1.84
25	0.087	4.82
50	0.27	15.0
100	1.00	55.6

* Calculated using the relationship in Edwards et al. 1978.

Some Finer Points About Gold and Silver

Gold is quite resistant to corrosion and does not tarnish like most other metals, such as iron and copper. Gold is one of the most malleable metals and can be flattened until it becomes translucent, and it can be stretched into a thin wire only 5 μm (micrometers) in diameter. Gold is quite resistant to corrosion and a very good conductor of electricity, thus finding use in electronic circuits. The symbol Au is used for gold, which has its origin in the Latin word *aurum*, meaning "glow of sunrise." The largest gold nugget in history was found in Australia in 1869, and it weighed 78 kg (2,508 oz).

Unlike gold, silver does tarnish, so silver jewelry and utensils must be periodically cleaned to remove the tarnish. Tarnish is usually the result of silver reacting with sulfur, as may be present in water, soaps, and foods like onions and eggs. Of all the elements, silver is the most electrically conductive, has the highest thermal conductivity, and is the most reflective. Exceptionally high-quality mirrors can be constructed from silver.

The symbol Ag is used for silver, which has its origin in the Latin word *argentum*, simply meaning "silver." The country of Argentina derives its name from silver as does the river Rio de la Plata (*plata* is the Spanish word for silver).

Source: Parsons and Dixon 2013; Gray 2009

SIMPLE CYANIDES

The compounds sodium cyanide (NaCN), potassium cyanide (KCN), and calcium cyanide ($Ca(CN)_2$) are often referred to as "simple cyanides." These are solid cyanide compounds with high solubilities in water. In solution, they do not exist in unionized form, except possibly in highly concentrated solutions. When mixed with water, these simple cyanides readily dissolve and dissociate into free cyanide and the corresponding cations of Na^+, K^+, and Ca^{2+}. Of these compounds, NaCN is the most important in gold and silver leaching because it is supplied in industrial quantities. Chemically, however, calcium cyanide and potassium cyanide have the same effectiveness for leaching gold and silver because the active constituent is CN^- in these compounds.

UNIT CONVERSIONS

It is occasionally necessary to convert concentration units for some basic cyanide compounds, and the molecular weights needed for such conversions are listed in Table 2.3.

TABLE 2.3 Molecular weights for basic cyanide compounds

Compound	Molecular Weight
CN^-	26.017
HCN	27.025
NaCN	49.007

The most common is the conversion between CN^- and NaCN, for which the following relationship applies:

$$[\text{mg/L as NaCN}] = [\text{mg/L as CN}] \times 1.884$$

A word of caution is offered about the concentration units used when the results from a silver nitrate titration are reported. Leach plant operators often prefer units of milligrams per liter or grams per metric ton as NaCN, as opposed to milligrams per liter or grams per metric ton as CN. This is because concentrations expressed in terms of NaCN are more easily related to NaCN consumed in the process. However, this practice can vary from plant to plant, and sometimes there is confusion about which units are actually being used. Be sure to confirm with the laboratory performing the titration, as the difference is quite significant. Cyanide concentrations reported by a laboratory should always include the relevant units of measurement. The standard practice for commercial analytical laboratories is to utilize units of milligrams per liter as CN when reporting concentrations of free, weak acid dissociable [WAD], and total cyanide.

79
Au
Gold
196.966569

3

Gold

The common oxidation states for gold are +1 and +3, but in cyanide solution, only the +1 state exists. The overall reaction for leaching of gold with cyanide in aerated alkaline solution is shown in the following equation in what is commonly called *Elsner's equation*. In this reaction, elemental gold is oxidized from Au^0 metal to Au^+, with dissolved oxygen serving as the oxidant. Gold then binds with cyanide to yield the soluble and very stable complex $Au(CN)_2^-$.

$$4Au^0 + 8CN^- + O_2 + 2H_2O \rightarrow 4Au(CN)_2^- + 4OH^-$$

In the absence of either dissolved oxygen (or another oxidant) or cyanide, this leach reaction will not occur. In air-saturated solutions, a minimum free cyanide concentration of about 50 mg/L is needed to prevent cyanide from kinetically limiting the rate of leaching (Yannopoulos 1991; van der Merwe and Breuer 2013).

Stability constants for the common gold-cyanide complexes are shown in Table 3.1.

TABLE 3.1 Stability constants for gold-cyanide complexes

Equilibrium Reaction	log β (25°C)	Reference
$Au^+ + CN^- \rightleftharpoons AuCN_{(aq)}$	18	Flynn and McGill 1995
$Au^+ + 2CN^- \rightleftharpoons Au(CN)_2^-$	38.92	NBS 1982*
$Au^+ + CN^- + OH^- \rightleftharpoons Au(OH)(CN)^-$	30	Flynn and McGill 1995
$Au^+ + CN^- \rightleftharpoons AuCN_{(s)}$	32	Flynn and McGill 1995

* Calculated by the author using STABCAL (Montana Tech, n.d.) and NBS (1982).

Gold primarily occurs in ores in native form and as an alloy with silver (electrum) but occasionally gold-tellurides are encountered. Gold is frequently associated with sulfidic minerals such as pyrite, marcasite, pyrrhotite, arsenopyrite, chalcopyrite, galena, and sphalerite. These sulfidic minerals can complicate gold leaching and recovery, as discussed in later chapters.

An Eh-pH diagram for the aqueous gold-cyanide system is shown in Figure 3.1. As indicated, $Au(CN)_2^-$ is the predominant complex in oxidized leach solution, with elemental gold Au^0 metal being predominant only under reducing conditions. Such reducing conditions are well known in electrowinning and zinc cementation (Merrill–Crowe) circuits. The high stability of the $Au(CN)_2^-$ complex makes this equilibrium chemistry relatively simple and shows that the complex is stable over a wide range of Eh and pH conditions. For practical purposes, other soluble gold complexes do not exist in cyanide leach solutions.

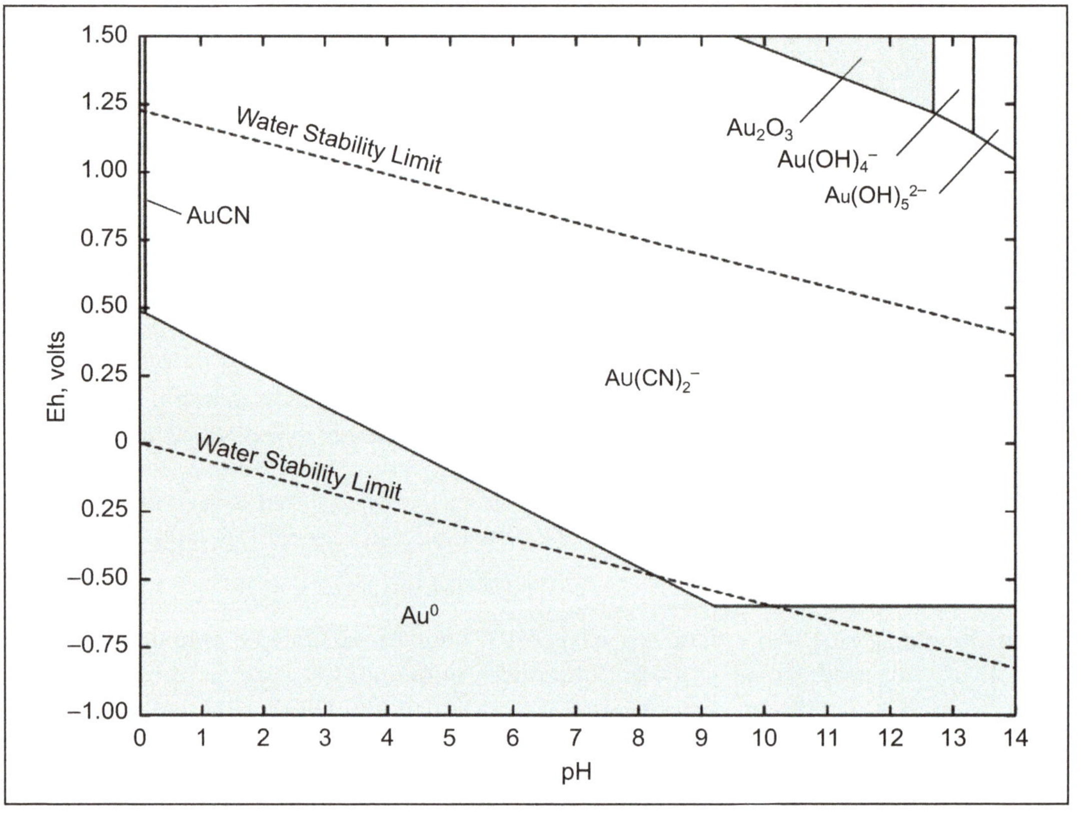

FIGURE 3.1 Eh-pH diagram for Au-CN-H_2O system (a[Au] $10^{-4.6}$, a[CN] $10^{-2.4}$, 25°C)

Why Troy Ounces?

Gold and silver are bought and sold worldwide using the unit of weight *troy ounces*. There are 12 troy ounces in 1 troy pound. This is different from the *avoirdupois* system of measure, in which there are 16 ounces in 1 pound. The troy system is believed to have its origin long ago in the city of Troyes in France, which was an important commercial trade center and benefited from a standardized system of weights and measures. The early use of the troy ounce and the momentum of history now give us the weight that is used today for trading gold; silver; and the other precious metals of platinum, palladium, iridium, osmium, rhodium, and ruthenium. At some point, perhaps, the metric system may prevail, but that is not likely to happen any time soon.

12 troy ounces = 1 troy pound
14.583333 troy ounces = 1 lb (pound)
1 troy ounce = 31.1034768 g (grams)
1 troy oz/ton (ounce/ton) = 34.2857142857147 g/t (grams/metric ton)

Source: Hallock and Herbert 1906

Golden Superlatives

The word "golden" is often used as a superlative to describe something as being highly valuable, important, or superb. Examples include golden rule, golden parachute, golden anniversary, golden age, golden opportunity, golden years, and so forth. Gold has a high monetary value and so this word is used to ascribe value to the things of most importance to us.

And the words of the Lord are flawless,
like silver purified in a crucible,
like gold refined seven times. (Psalm 12:6, NIV)

A pH-distribution diagram for the aqueous gold-cyanide system is rather uninteresting, given that only the $Au(CN)_2^-$ complex exists at any Eh-pH conditions of practical interest in cyanidation. At a very low pH, gold can precipitate as the solid $AuCN_{(s)}$ (Flynn and McGill 1995; Ford-Smith 1964). There is also a possibility the complex $Au(CN)_3^{2-}$ could form with very high levels of cyanide in solution.

The $Au(CN)_2^-$ complex has a high affinity for adsorption onto activated carbon. The general order of preference for adsorption is shown in Figure 3.2 relative to other common metal-cyanide complexes. An important exception to the indicated preferences is the $Cu(CN)_2^-$ complex, which has a high affinity for adsorption onto activated carbon. This is further discussed in Chapter 8. In general, the dicyano metal complexes with 0 or −1 charge (e.g., Ag, Au, Cd, Hg, Cu) are more strongly adsorbed by carbon, while the metal complexes with more than two cyanide groups attached and with a more negative charge are less strongly adsorbed (Bailey 1987).

$Au(CN)_2^- > Hg(CN)_2^0 > Ag(CN)_2^- > Cu(CN)_3^{2-} > Zn(CN)_4^{2-} > Ni(CN)_4^{2-} > Fe(CN)_6^{4-}$

← Increasing Preference for Adsorption

Source: Marsden and House 2006

FIGURE 3.2 General preference for adsorption onto activated carbon

The World's Largest Gold Deposit

The world's oceans are estimated to contain an average of 4×10^{-6} mg/L of gold. This gold would likely be present as a soluble chloride complex, with gold in the Au^{3+} oxidation state and chloride acting as the ligand. The total volume of the world's oceans is about 1.33×10^9 km^3. Thus, the oceans hold about 5.32×10^{12} g of gold, or about 171 billion oz. At today's market price (first quarter of 2023) of US$1,900/oz, the value would be an astonishing US$325 trillion! Many have tried to recover this gold but to no avail so far. Perhaps the Bering Sea would be a good starting place, as the concentrations are highest there due to the gold-rich rivers in Alaska and Siberia.

Source: Parsons and Dixon 2013; Lide 2005; Charette and Smith 2010

47
Ag
Silver
107.8682

Silver

4

The stable oxidation state for silver is +1, and this is the state encountered in cyanide leach solutions. Leaching of silver with cyanide proceeds in the same general manner as for gold, as shown in the following reaction. In this reaction, elemental silver is oxidized from Ag^0 metal to Ag^+, with dissolved oxygen serving as the oxidant. Silver then binds with cyanide to yield the soluble complex $Ag(CN)_2^-$. Silver generally leaches more slowly than gold, frequently with lower overall extractions, and with higher cyanide concentrations required.

$$4Ag^0 + 8CN^- + O_2 + 2H_2O \rightarrow 4Ag(CN)_2^- + 4OH^-$$

Stability constants for the common silver-cyanide complexes are shown in Table 4.1.

TABLE 4.1 Stability constants for silver-cyanide complexes

Equilibrium Reaction	log β (25°C)	Reference
$Ag^+ + CN^- \rightleftharpoons AgCN_{(aq)}$	9.7	Flynn and McGill 1995
$Ag^+ + 2CN^- \rightleftharpoons Ag(CN)_2^-$	20.44	Izatt et al. 1967
$Ag^+ + 3CN^- \rightleftharpoons Ag(CN)_3^{2-}$	21.99	Izatt et al. 1967
$Ag^+ + 4CN^- \rightleftharpoons Ag(CN)_4^{3-}$	21.4	Flynn and McGill 1995
$Ag^+ + CN^- + OH^- \rightleftharpoons Ag(OH)(CN)^-$	13.45	NBS 1982*
$Ag^+ + CN^- \rightleftharpoons AgCN_{(s)}$	15.67	NIST 2004†
$2Ag^+ + 2CN^- \rightleftharpoons Ag_2(CN)_{2(s)}$	11.16	NIST 2004†

* Calculated by the author using STABCAL (Montana Tech, n.d.) and NBS (1982).

† Calculated by the author using STABCAL (Montana Tech, n.d.) and NIST (2004).

Silver occurs in ores in native form, as an alloy with gold (electrum), and as the sulfide mineral Ag_2S known as acanthite (or argentite). Acanthite is soluble in cyanide, but oxidation of the bound sulfide is also required for leaching to occur. This primarily yields thiosulfate ($S_2O_3^{2-}$) as the sulfur product, as shown in the following reaction (Luna and Lapidus 2000):

$$Ag_2S + 4CN^- + O_2 + \tfrac{1}{2}H_2O \rightarrow 2Ag(CN)_2^- + \tfrac{1}{2}S_2O_3^{2-} + OH^-$$

Leaching of silver from acanthite consumes about twice the amount of dissolved oxygen as does the leaching of elemental silver (not accounting for the potential impacts of other sulfide

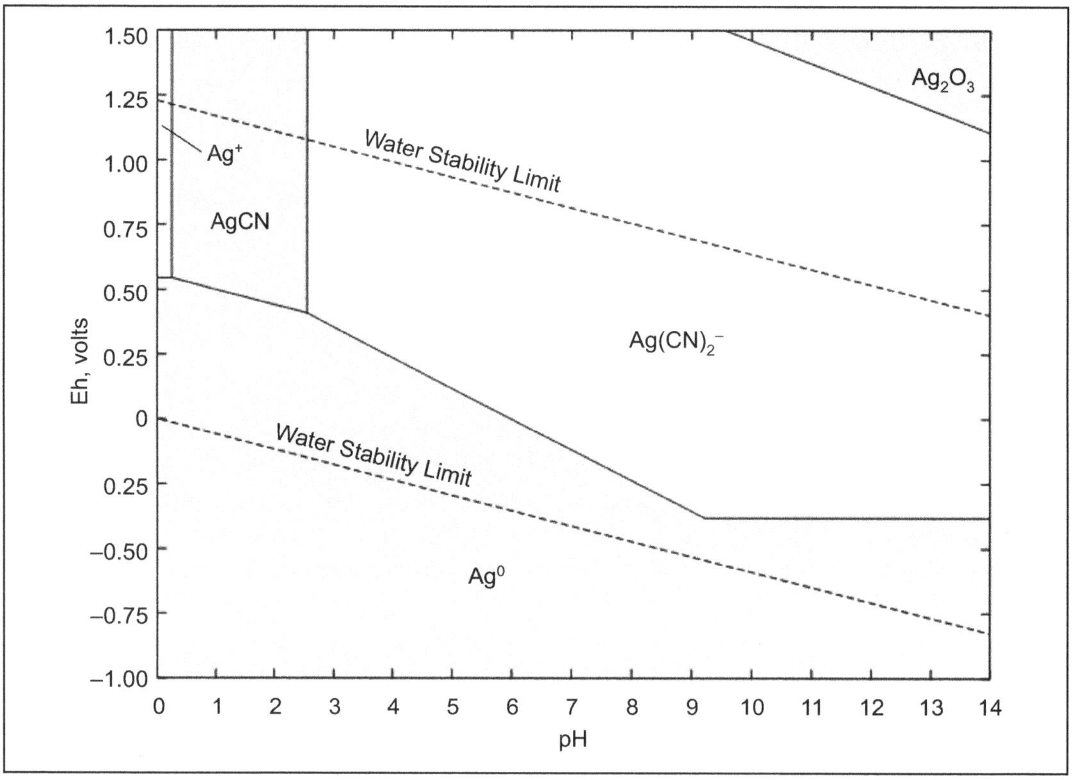

FIGURE 4.1 Eh-pH diagram for Ag-CN-H_2O system (a[Ag] $10^{-4.3}$, a[CN] $10^{-2.4}$, 25°C)

minerals in the ore). Lead nitrate has been employed when leaching acanthite to improve the kinetic leach rate and final extraction of silver (Deschênes et al. 2011).

An Eh-pH diagram for the aqueous silver-cyanide system is shown in Figure 4.1. The predominant silver complex is $Ag(CN)_2^-$, except at low pH where the solid $AgCN_{(s)}$ and the free silver cation Ag^+ occur. If thiocyanate is present in solution at low pH, $AgSCN_{(s)}$ may precipitate in preference to $AgCN_{(s)}$.

A pH-distribution diagram for the aqueous silver-cyanide system is shown in Figure 4.2. In the common pH range of 10–11 utilized for cyanide leaching, the complex $Ag(CN)_2^-$ predominates, with only a minor amount of the silver being present as $Ag(CN)_3^{2-}$ (~6% in this case). Only at a pH of less than about 3.5 do the complexes $AgCN_{(s)}$, $AgCN_{(aq)}$, and Ag^+ begin to appear. With higher cyanide concentrations, the relative proportions of $Ag(CN)_3^{2-}$ and $Ag(CN)_4^{3-}$ in solution increase, but $Ag(CN)_2^-$ still remains the predominant complex in most situations.

An Eh-pH diagram for the aqueous silver-cyanide-sulfide system is shown in Figure 4.3. In the presence of sulfide, a region of stability exists for Ag_2S solid. But as just noted, Ag_2S (acanthite) is soluble in cyanide solution provided that the sulfide portion of this mineral is oxidized. In the absence of dissolved oxygen (or another oxidant), Ag_2S remains insoluble. The overall leaching rate of Ag_2S is generally limited by the rate of oxidation of sulfide to form thiosulfate or another soluble sulfur compound.

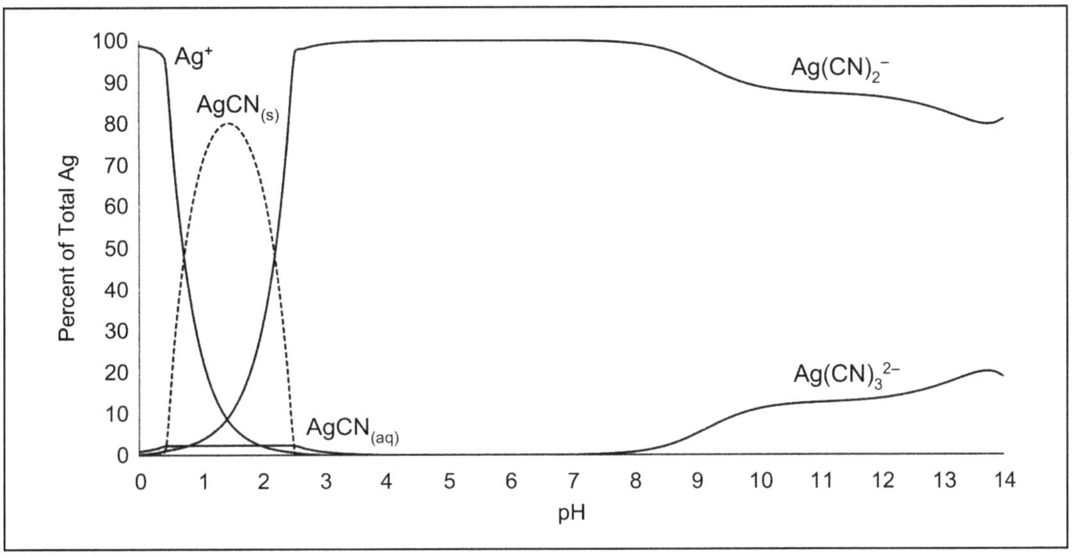

FIGURE 4.2 pH-distribution diagram for Ag-CN-H_2O system (Ag^+ 5 mg/L, CN^- 100 mg/L, 25°C)

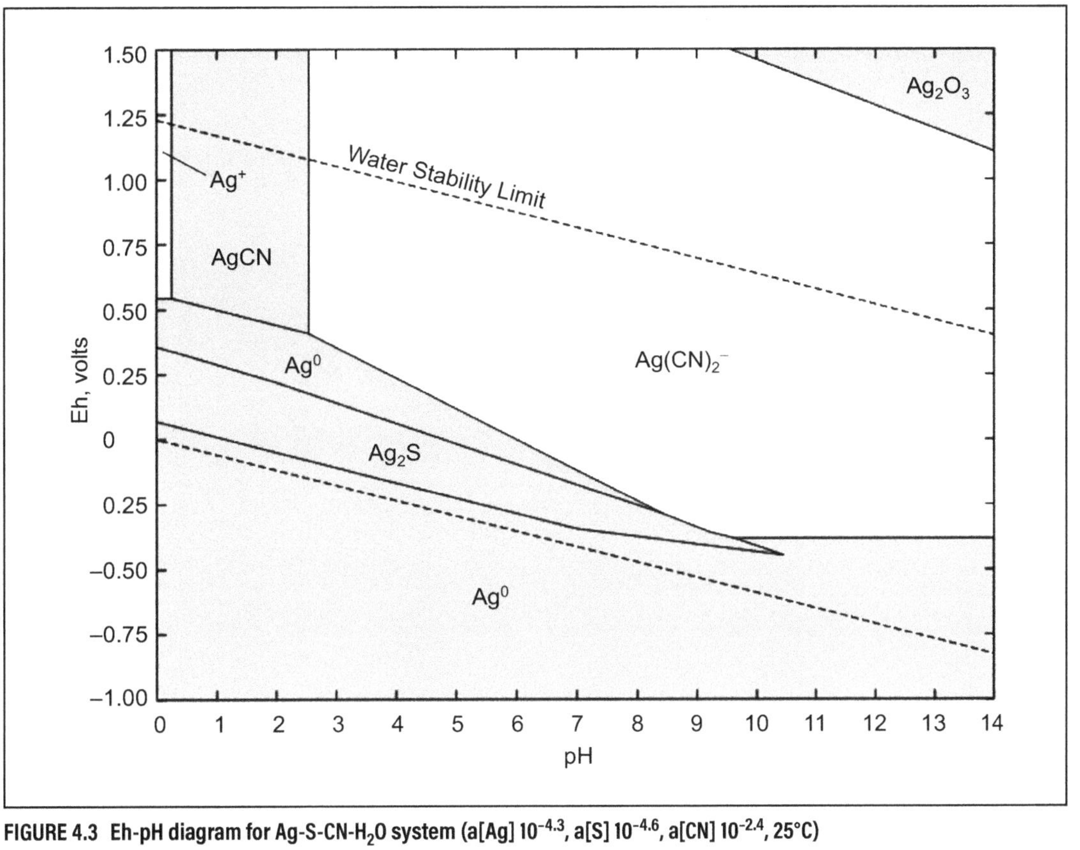

FIGURE 4.3 Eh-pH diagram for Ag-S-CN-H_2O system (a[Ag] $10^{-4.3}$, a[S] $10^{-4.6}$, a[CN] $10^{-2.4}$, 25°C)

26
Fe
Iron
55.845

Iron

5

The common oxidation states for iron are +2 and +3, and in cyanide leach solution, both forms may be present. The iron-cyanide complexes are well known to be quite stable and are essentially inert in regard to their ability to leach gold or silver. Stability constants for the common iron-cyanide complexes are shown in Table 5.1.

TABLE 5.1 Stability constants for iron-cyanide complexes*

Equilibrium Reaction	log β (25°C)	Reference
$Fe^{2+} + 6CN^- \rightleftharpoons Fe(CN)_6^{4-}$	45.63	Broderius 1973 †
$Fe^{2+} + 6CN^- + H^+ \rightleftharpoons HFe(CN)_6^{3-}$	49.79	NBS 1982†
$Fe^{2+} + 6CN^- + 2H^+ \rightleftharpoons H_2Fe(CN)_6^{2-}$	52.02	NBS 1982†
$Fe^{3+} + 6CN^- \rightleftharpoons Fe(CN)_6^{3-}$	52.61	NBS 1982†; Broderius 1973
$3Fe^{2+} + 2Fe^{3+} + 12CN^- \rightleftharpoons Fe_3(Fe(CN)_6)_{2(s)}$‡	177.4	Dzombak et al. 2006
$3Fe^{2+} + 4Fe^{3+} + 18CN^- \rightleftharpoons Fe_4(Fe(CN)_6)_{3(s)}$§	263.6	Dzombak et al. 2006
$2Cu^{2+} + Fe^{2+} + 6CN^- \rightleftharpoons Cu_2Fe(CN)_{6(s)}$	61.31	NBS 1982†

* Stability constants for the complexes $Fe(CN)_6^{4-}$ and $Fe(CN)_6^{3-}$ vary widely in published literature.
† Calculated by the author using STABCAL (Montana Tech, n.d.) and NBS (1982).
‡ Turnbull's blue.
§ Prussian blue.

The primary source of iron in leach solution is through the reaction of cyanide with various iron-sulfide minerals. Most commonly, these minerals include pyrite, marcasite, chalcopyrite, arsenopyrite, and pyrrhotite. Iron in these minerals is present either in the ferrous (Fe^{2+}) or ferric (Fe^{3+}) oxidation state, as shown in Table 5.2. However, many different reactions can occur between cyanide and the iron-sulfide minerals, some of which lead to the formation of various soluble and insoluble iron-cyanide complexes. The minerals listed in Table 5.2 all have iron in the Fe^{2+} oxidation state, except for chalcopyrite. There is some disagreement in the body of literature regarding this point, with some advocating oxidation states of Cu^{2+} and Fe^{2+} in chalcopyrite.

Fortunately, pyrite, chalcopyrite, and arsenopyrite generally have low reactivities toward cyanide. But pyrrhotite and marcasite are more reactive. The author has encountered leach circuits with dissolved iron concentrations exceeding 300 mg/L as a result of pyrrhotite and marcasite reacting with cyanide. Normally, however, iron concentrations in leach solutions are

TABLE 5.2 Oxidation states in common iron-sulfide minerals

Mineral*	Formula	Oxidation States†
Pyrite	FeS_2	Fe^{2+} S_2^{2-}
Marcasite	FeS_2	Fe^{2+} S_2^{2-}
Chalcopyrite	$CuFeS_2$	Cu^+ Fe^{3+} S^{2-}
Arsenopyrite	FeAsS	Fe^{2+} As^- S^-
Pyrrhotite	$Fe_{(1-n)}S$ (n = 0–0.2)	Fe^{2+} S^{2-} (n = 0)

* Marcasite and pyrite are polymorphs and have the same chemical formula. The crystal structure of pyrite is cubic and the crystal structure of marcasite is the less stable orthorhombic.

† The oxidation states of Cu^+ and Fe^{3+} in chalcopyrite were sourced from Goh et al. 2006.

relatively low, often in the range of about 1–50 mg/L. The general tendencies of some iron minerals to be solubilized by cyanide are listed in Table 5.3.

TABLE 5.3 Solubility of some iron minerals by cyanide

Mineral	Formula	Solubility
Pyrite	FeS_2	Sparingly
Pyrrhotite	$Fe_{(1-n)}S$ (n = 0–0.2)	Readily
Hematite	Fe_2O_3	Insoluble
Magnetite	Fe_3O_4	Insoluble
Goethite	FeOOH	Insoluble
Siderite	$FeCO_3$	Insoluble

Source: Marsden and House 2006; Habashi 1967

An Eh-pH diagram for the aqueous iron-cyanide system is shown in Figure 5.1. In alkaline solution, the iron-cyanide complexes of $Fe(CN)_6^{4-}$ (ferrous) and $Fe(CN)_6^{3-}$ (ferric) are the predominant forms, but $Fe(CN)_6^{4-}$ has a larger overall region of stability. It is reported that the presence of free cyanide will slowly reduce $Fe(CN)_6^{3-}$ to $Fe(CN)_6^{4-}$ and that the reaction is strongly catalyzed by copper (Flynn and McGill 1995). The ratio of 6 moles of cyanide per 1 mole of iron in these complexes indicates each 1 mg/L of Fe in solution is associated with 2.8 mg/L of CN. This bound cyanide is inert toward leaching of gold and silver because of the high stability of the iron-cyanide complexes. Thus, the reactions of cyanide with iron have the potential to result in high consumptions of cyanide. The formation of iron-cyanide represents an irreversible loss of usable cyanide (i.e., iron acts as a "cyanicide").

Oxidation and reduction of the iron-cyanide complexes is kinetically slow at alkaline pH, so equilibrium based on particular Eh conditions may not always be achieved. At an alkaline

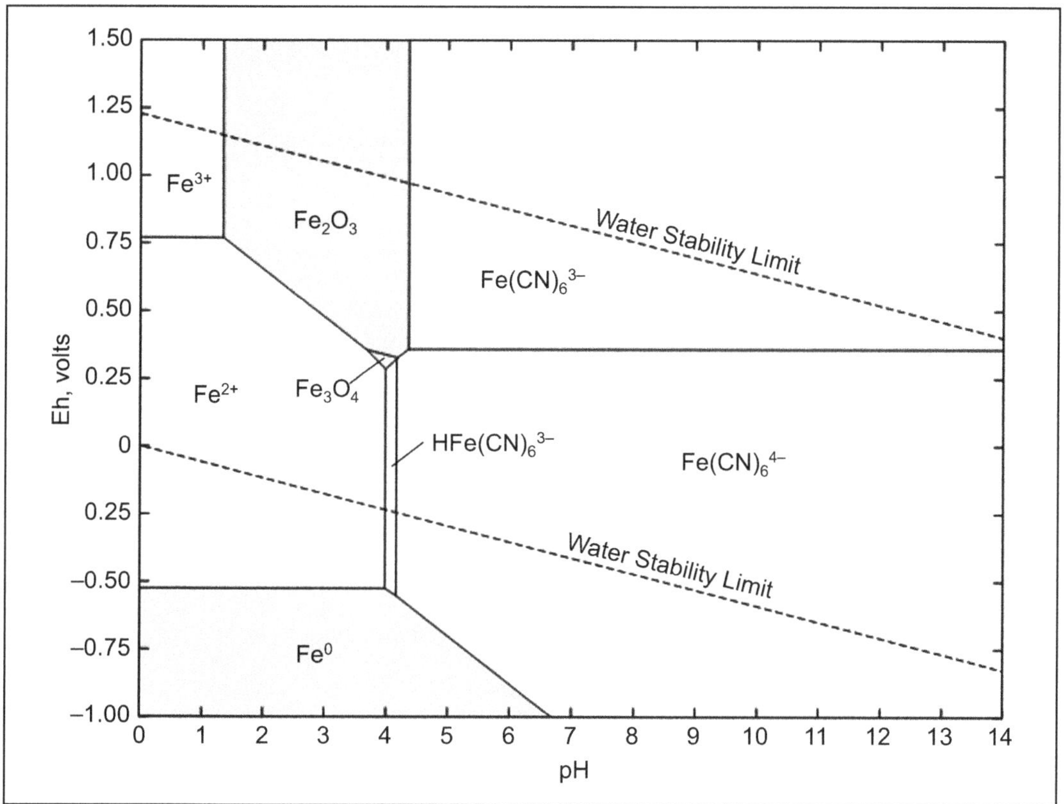

FIGURE 5.1 Eh-pH diagram for Fe-CN-H_2O system (Prussian blue and Turnbull's blue not considered) (a[Fe] $10^{-4.0}$, a[CN] $10^{-2.4}$, 25°C)

pH, oxidation of ferrous-cyanide to ferric-cyanide requires a strong oxidant. At neutral to acidic pH conditions, ferrous-cyanide is readily oxidized to ferric-cyanide.*

Figure 5.1 suggests the iron-cyanides will dissociate to yield free Fe^{2+} or Fe^{3+} under acidic conditions. In practice, the dissociation reactions are slow and only occur under relatively aggressive conditions for pH and temperature. However, the iron-cyanides are photolytically active and can dissociate to yield free cyanide when exposed to ultraviolet light.

Ferrous iron (Fe^{2+}) reacts fairly rapidly with cyanide to form $Fe(CN)_6^{4-}$. By contrast, ferric iron reacts more slowly with cyanide to form $Fe(CN)_6^{3-}$. This characteristic is a reason pre-aeration steps can be helpful upstream of a leach circuit before the addition of cyanide. Pre-aeration oxidizes a portion of ferrous iron to ferric iron on mineral surfaces, which then renders the minerals less reactive toward cyanide.

The stability region for Fe_2O_3 (hematite) shown in Figure 5.1 is often identified as goethite or another iron oxide/hydroxide solid on Eh-pH diagrams. In practice, a number of iron

* A common oxidation reduction potential (ORP) calibration standard is based on this behavior. ZoBell's solution is an ORP standard and consists of a mixture of potassium ferrocyanide and potassium ferricyanide buffered at pH 7.

The Richest Hill on Earth

Visitors to Butte, Montana (United States) are likely informed it is the home of the "richest hill on Earth." The total amounts of metals produced from the richest hill on Earth over the period 1880–2005 are reported to be:

Gold	2,922,446 oz
Silver	715,976,036 oz
Copper	21,635,294,319 lb
Zinc	4,909,202,540 lb
Manganese	3,702,787,341 lb
Lead	854,797,405 lb
Molybdenum	197,007,798 lb

If these metals were valued according to 2021 pricing, as reported in the *Mineral Commodity Summaries* of the U.S. Geological Survey, the total would exceed US$125 billion. The richest hill is still being mined today and is home to a unique people, eclectic history, and stunning scenery. Part of my honeymoon was spent in Butte, to the continued amusement of my wife of 32 years.

Source: Czehura 2006

oxide/hydroxide solids can form in this region, depending on the conditions under which the solids are formed. The general order of stability of the common iron oxide/hydroxide solids is shown in the following (Butler 1998):

$$\text{hematite } (Fe_2O_3) > \text{goethite } (FeOOH) > \text{ferrihydrite } (Fe(OH)_{3(am)})$$

It seems when ferric iron precipitates at ambient temperatures, ferrihydrite is first formed. Then as the precipitate ages, ferrihydrite may transition to the more stable compounds of goethite or hematite, although magnetite (Fe_3O_4) is also a possibility. Ferrihydrite is often adopted as the stable solid when considering iron precipitation systems (Butler 1998; Bethke 2008; Krauskopf 1967; Greenwood and Earnshaw 1984). The actual solids formed depend on factors such as temperature and aging of the solids after their initial formation. It appears that at lower temperatures, ferrihydrite is favored; at moderate temperatures, goethite is favored; while at elevated temperatures, hematite is favored.

Formation of the solid compounds of Prussian blue ($Fe_4[Fe(CN)_6]_3$) and Turnbull's blue ($Fe_3[Fe(CN)_6]_2$) were suppressed when generating the Eh-pH diagram for iron-cyanide (Figure 5.1). These solid compounds can form under neutral to acidic pH conditions, provided iron in the ferrous (Fe^{2+}) or ferric (Fe^{3+}) state is available (e.g., through the dissolution of iron minerals, dissolution of grinding media, or by intentionally dosing iron-containing reagents). The Eh-pH diagram for the case where the formations of Prussian blue and Turnbull's blue are allowed is shown in Figure 5.2. As indicated, a pH of less than about 7 favors their formation, with Prussian blue forming under more oxidizing conditions and Turnbull's blue forming

Cyanide and Blue Jeans

The first purely synthetic pigment was made in the 18th century from cyanide, and it was called *Prussian blue*. The method of preparation was a guarded secret because of the profitable business generated for dying textiles and tinting paints. The method of its preparation remained secret for only about 20 years, after which it came into widespread production and use in Europe. The earliest painting with Prussian blue dates from 1709 in the *Entombment of Christ* by Dutch painter Pieter van der Werff. Beautifully colored dyes and pigments have been made from various iron-cyanide compounds, including Prussian blue, Turnbull's blue, and Berlin green. The colors of maroon, white, yellow, orange-yellow, red-brown, brown, and red are also possible. Prussian blue is still used as a dye for textiles but is sold under the trade name Iron Blue. Prussian blue also has medical uses for treating thallium and cesium poisoning, as it readily binds with these metals, which may be radioactive and are highly toxic. In fact, the World Health Organization has Prussian blue on its list of essential medicines as a poisoning antidote.

Source: Ford-Smith 1964; Kraft 2008; American Cyanamid 1953; WHO 2013

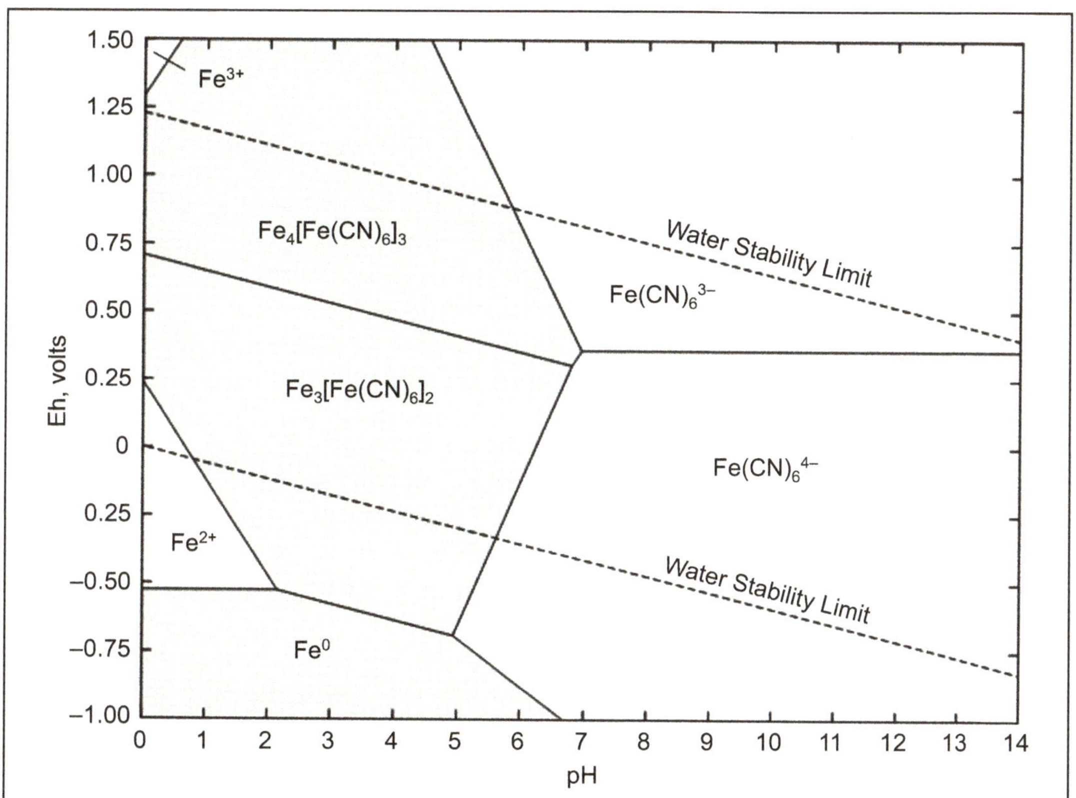

FIGURE 5.2 Eh-pH diagram for Fe-CN-H_2O system (Prussian blue and Turnbull's blue considered) (a[Fe] $10^{-4.0}$, a[CN] $10^{-2.4}$, 25°C)

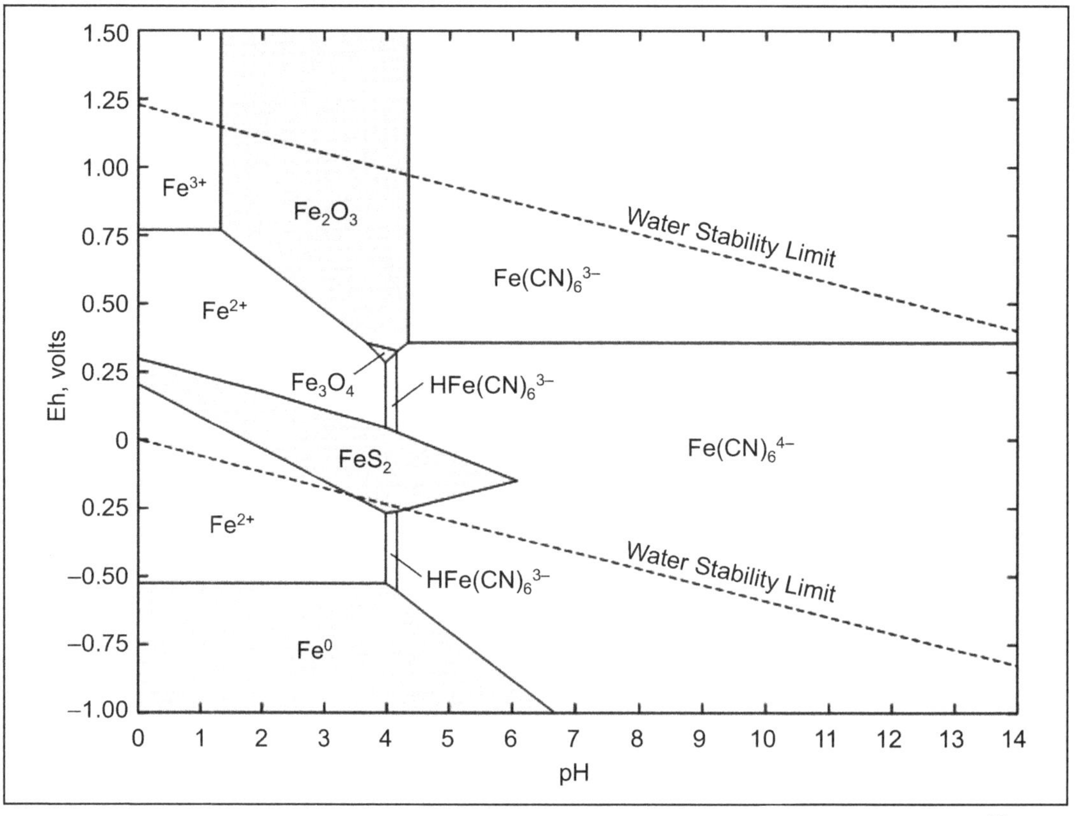

FIGURE 5.3 Eh-pH diagram for Fe-S-CN-H_2O system (Prussian blue and Turnbull's blue not considered) (a[Fe] $10^{-4.0}$, a[S] $10^{-3.7}$, a[CN] $10^{-2.4}$, 25°C)

under more reducing conditions. At alkaline pH and in the absence of light, Prussian blue and Turnbull's blue solids can redissolve at a slow rate (Dzombak et al. 2006).

Within a heap leach, it is possible that localized zones with a low pH could result in the precipitation of iron-cyanide compounds (i.e., the behavior indicated in Figure 5.2 at lower pH). This likely depends on the concentrations of cyanide within these zones and the presence of iron from sulfide minerals.

An Eh-pH diagram for the aqueous iron-cyanide-sulfide system is shown in Figure 5.3. In the presence of sulfide, a relatively small stability region develops for FeS_2 solid (pyrite) under acidic and reducing conditions. Others have reported an additional small region of stability in this same area for FeS (pyrrhotite) (Osseo-Asare et al. 1984). This Eh-pH diagram suggests pyrite is soluble in cyanide solution at alkaline pH, but in practice, pyrite is fairly nonreactive toward cyanide (Table 5.3).

A pH-distribution diagram for the aqueous iron-cyanide system is not presented, given that only the $Fe(CN)_6^{4-}$ and $Fe(CN)_6^{3-}$ complexes will exist in solution at any Eh-pH conditions of practical interest in cyanidation.

27
Co
Cobalt
58.933194

6 Cobalt

The common oxidation states for cobalt are +2 and +3. In aqueous solution, and in the absence of complexing ligands, Co^{3+} is unstable and is such a strong oxidant that it reduces to Co^{2+} while decomposing water to form oxygen gas (Baes and Mesmer 1976). But in alkaline cyanide solutions, Co^{3+} is the more stable form, and Co^{2+} is readily oxidized to Co^{3+}. Stability constants for the common cobalt-cyanide complexes are shown in Table 6.1.

TABLE 6.1 Stability constants for cobalt-cyanide complexes*

Equilibrium Reaction	log β (25°C)	Reference
$Co^{2+} + 3CN^- \rightleftharpoons Co(CN)_3^-$	14.5	NIST 2004†
$Co^{2+} + 5CN^- \rightleftharpoons Co(CN)_5^{3-}$	23	NIST 2004†
$Co^{2+} + 6CN^- \rightleftharpoons Co(CN)_6^{4-}$	19.1	Lorösch 2001
$Co^{3+} + 5CN^- \rightleftharpoons Co(CN)_5^{2-}$	62	Flynn and McGill 1995
$Co^{3+} + 6CN^- \rightleftharpoons Co(CN)_6^{3-}$	70.04	NIST 2004†
$Co^{2+} + 2CN^- \rightleftharpoons Co(CN)_{2(s)}$	18	Flynn and McGill 1995

* There is some uncertainty about whether $Co(CN)_6^{4-}$ actually exists in cyanide solution. Sharpe (1976) reports that "$Co(CN)_6^{4-}$ is not present in analytically detectable amount in a solution containing $Co(CN)_5^{-3}$ and CN^-, but it appears to be a kinetic intermediate in some reactions of such a solution." For this handbook, the stability constant reported by Lorösch (2001) was considered.
† Calculated by the author using STABCAL (Montana Tech, n.d.) and NIST (2004).

Cobalt is occasionally present in gold and silver ores but typically at grades that do not result in elevated concentrations in leach solution. Common cobalt minerals include cobaltite (CoAsS), smaltite ($CoAs_3$), and linnaeite (Co_3S_4).

An Eh-pH diagram for the aqueous cobalt-cyanide system is shown in Figure 6.1. The complex $Co(CN)_6^{3-}$ has a large region of predominance and is essentially the only form of cobalt-cyanide that exists in a cyanidation circuit. Figure 6.1 suggests $Co(CN)_6^{3-}$ dissociates under acidic conditions to yield Co^{2+} and free cyanide. But in practice, this dissociation reaction is slow, such that for practical process purposes it does not occur. The $Co(CN)_6^{3-}$ complex can be considered refractory to complete dissociation, even in the presence of strong oxidants, sulfide, and extreme pH levels. Cobalt-cyanide is photolytically active but even with exposure

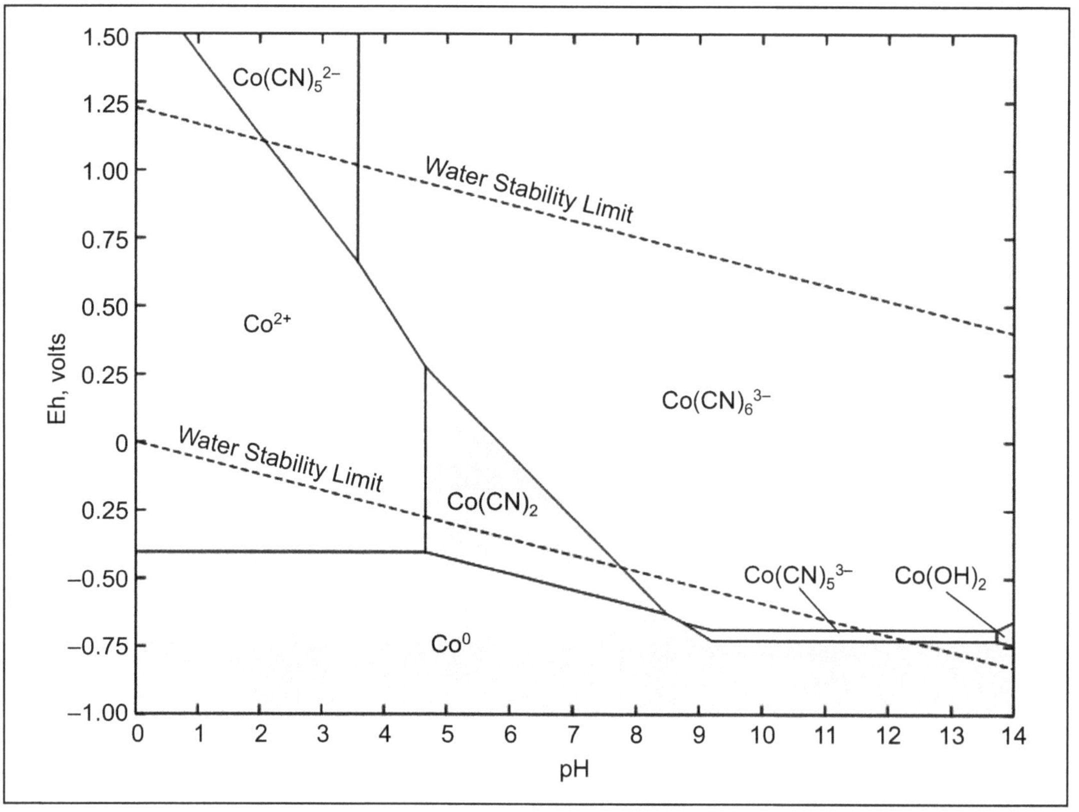

FIGURE 6.1 Eh-pH diagram for Co-CN-H_2O system (a[Co] $10^{-4.1}$, a[CN] $10^{-2.4}$, 25°C)

to ultraviolet light, the dissociation to yield free cyanide is minor within the time frames of interest in cyanidation.*

A pH-distribution diagram for the cobalt-cyanide system is shown in Figure 6.2. In the pH range of 10–11 commonly utilized for cyanide leaching, the complex $Co(CN)_6^{3-}$ predominates, as is expected from the Eh-pH diagram (Figure 6.1). As the pH becomes more acidic, the complex $Co(CN)_5^{2-}$ forms, although the kinetics for formation of this complex are likely slow. Although free cobalt Co^{2+} is thermodynamically predicted at low pH, this does not actually occur in practice, as just discussed.

An Eh-pH diagram for the aqueous cobalt-cyanide-sulfide system is shown in Figure 6.3. In the presence of sulfide, a relatively large region of stability exists for cobalt sulfide (CoS) solid under reducing conditions. In an aerated and alkaline cyanide solution, a portion of sulfide in the CoS solid is likely oxidized and cobalt simultaneously dissolved into solution by cyanide to form $Co(CN)_6^{3-}$. As noted earlier, free cobalt Co^{2+} is thermodynamically predicted at low pH, but this does not actually occur in practice to any notable extent.

---

* As discussed in Chapter 14 (see the "WAD and Total Cyanide" section and Table 14.2), the analytical recovery of cyanide from cobalt-cyanide is about 15% when using ultraviolet light irradiation with a total cyanide method. This closely corresponds to the loss of 1 mole of cyanide from $Co(CN)_6^{3-}$ to form $Co(CN)_5^{2-}$. The removal of 1 mole of cyanide from $Co(CN)_6^{3-}$ corresponds to about 17% of the bound cyanide.

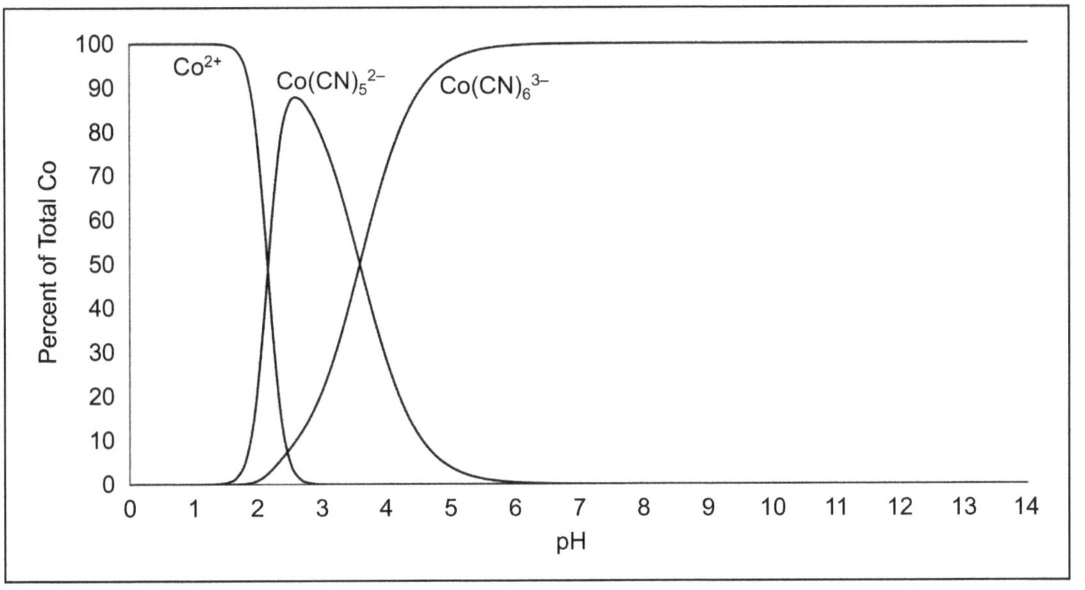

FIGURE 6.2 pH-distribution diagram for Co-CN-H_2O system (Co^{3+} 5 mg/L, CN^- 100 mg/L, 25°C, pO_2 0.2 atm) (Note: Formation of Co^{2+} is unlikely to occur in practice.)

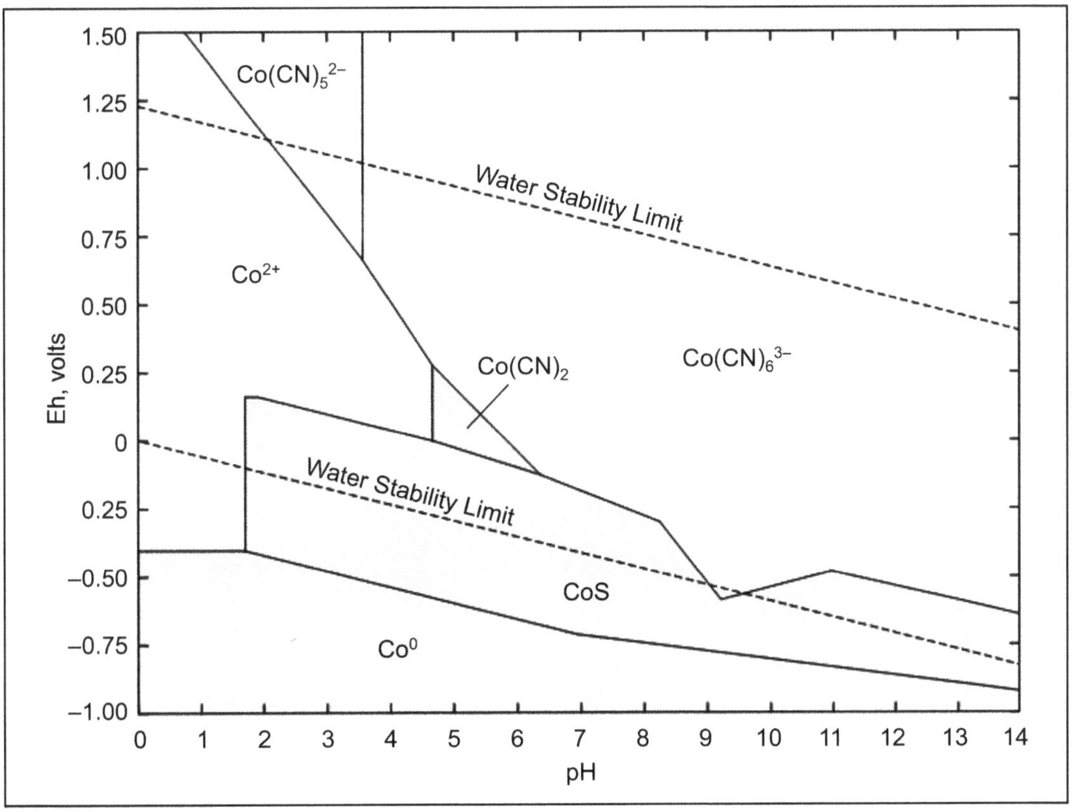

FIGURE 6.3 Eh-pH diagram for Co-S-CN-H_2O system (a[Co] $10^{-4.1}$, a[S] $10^{-4.1}$, a[CN] $10^{-2.4}$, 25°C)

28
Ni
Nickel
58.6934

Nickel

7

The stable oxidation state for nickel is +2 and this is the state encountered in cyanide leach solutions. Nickel is generally found at low levels in cyanide leach solutions because it is usually only present at trace levels in gold and silver ores, primarily in association with sulfide minerals. A common nickel-sulfide mineral is pentlandite with the general formula $(Ni,Fe)_9S_8$.

Stability constants for the common nickel-cyanide complexes are shown in Table 7.1.

TABLE 7.1 Stability constants for nickel-cyanide complexes

Equilibrium Reaction	log β (25°C)	Reference
$Ni^{2+} + 4CN^- \rightleftharpoons Ni(CN)_4^{2-}$	30.12	NBS 1982*
$Ni^{2+} + 5CN^- \rightleftharpoons Ni(CN)_5^{3-}$	28.5	Gamsjäger et al. 2005
$Ni^{2+} + 4CN^- + H^+ \rightleftharpoons NiH(CN)_4^-$	36.2	NIST 2004†
$Ni^{2+} + 4CN^- + 2H^+ \rightleftharpoons NiH_2(CN)_4^0$	40.93	NIST 2004†
$Ni^{2+} + 4CN^- + 3H^+ \rightleftharpoons NiH_3(CN)_4^+$	43.53	NIST 2004†
$Ni^{2+} + 2CN^- \rightleftharpoons Ni(CN)_{2(s)}$	19.5	Flynn and McGill 1995

* Calculated by the author using STABCAL (Montana Tech, n.d.) and NBS (1982).
† Calculated by the author using STABCAL (Montana Tech, n.d.) and NIST (2004).

An Eh-pH diagram for the aqueous nickel-cyanide system is shown in Figure 7.1. The predominant form of nickel in leach solution is $Ni(CN)_4^{2-}$ within quite a large region of Eh-pH stability. Thermodynamically, the complexes $NiH(CN)_4^-$ and $NiH_2(CN)_4^0$ are predicted to dissociate in acid solution to yield Ni^{2+} and free cyanide. But the kinetics of this dissociation at room temperature are slow, and in practice, the dissociation does not occur at a meaningful rate. Acidic dissociation of the nickel-cyanide complexes requires a strong competing ligand or elevated temperature. For the analysis of weak acid dissociable (WAD) cyanide using membrane separation techniques at room temperature, a polyamine compound is added as the competing ligand to complex with nickel and release free cyanide.

A pH-distribution diagram for the nickel-cyanide system is shown in Figure 7.2. In the pH range of 10–11 commonly utilized for cyanide leaching, the complex $Ni(CN)_4^{2-}$ predominates, as is expected from the Eh-pH diagram (Figure 7.1). As the pH becomes more acidic, the complexes $NiH(CN)_4^-$, $NiH_2(CN)_4^0$, and $NiH_3(CN)_4^+$ sequentially form, although the kinetics for the formation of these complexes are likely slow. Free nickel Ni^{2+} is thermodynamically

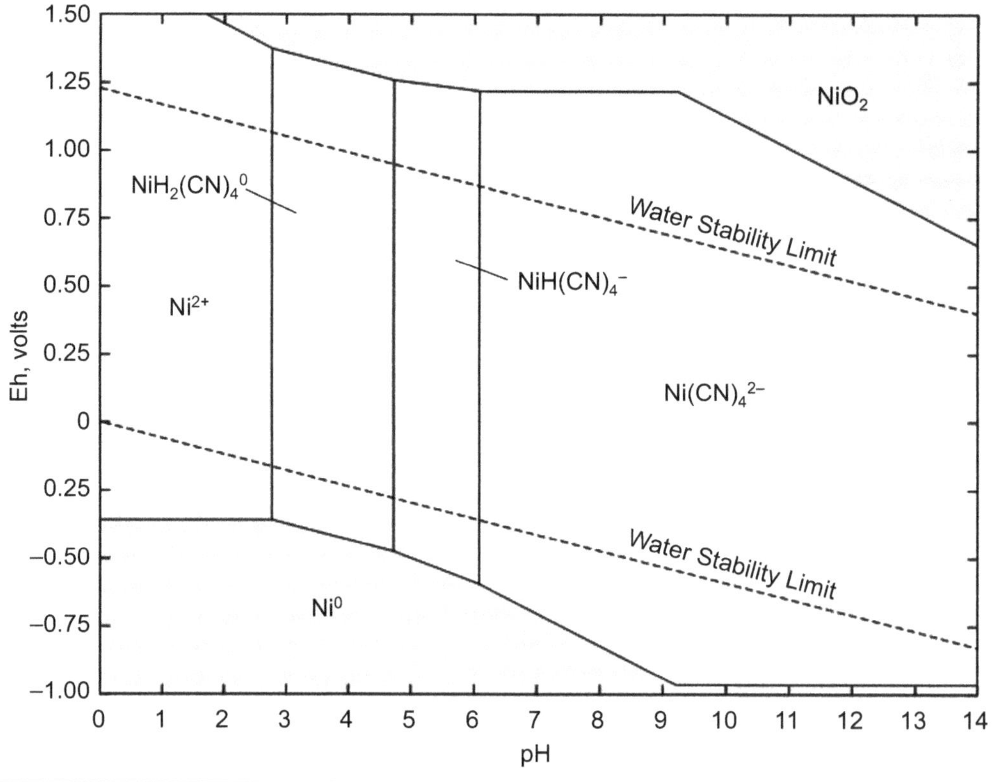

FIGURE 7.1 Eh-pH diagram for Ni-CN-H_2O system (a[Ni] $10^{-4.1}$, a[CN] $10^{-2.4}$, 25°C)

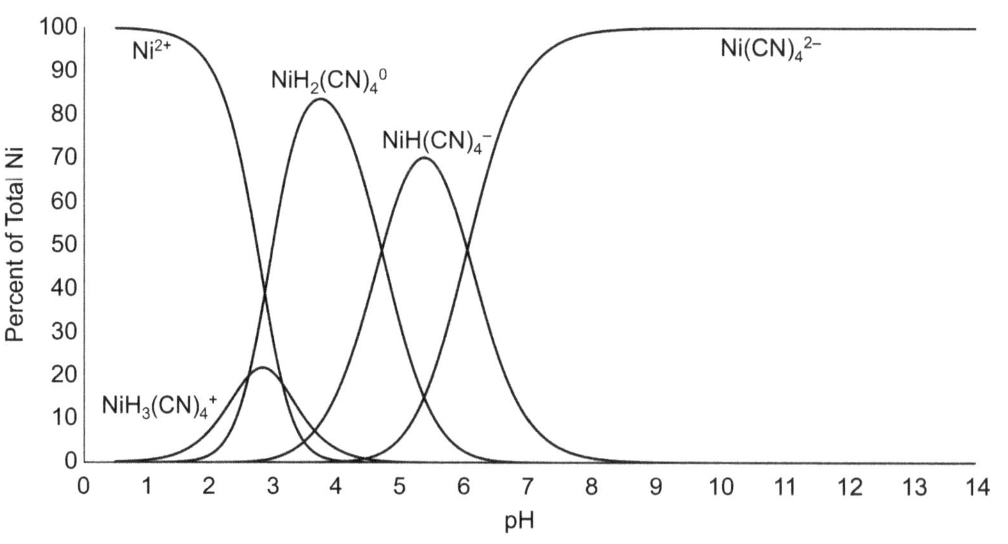

FIGURE 7.2 pH-distribution diagram for Ni-CN-H_2O system (Ni^{2+} 5 mg/L, CN^- 100 mg/L, 25°C)

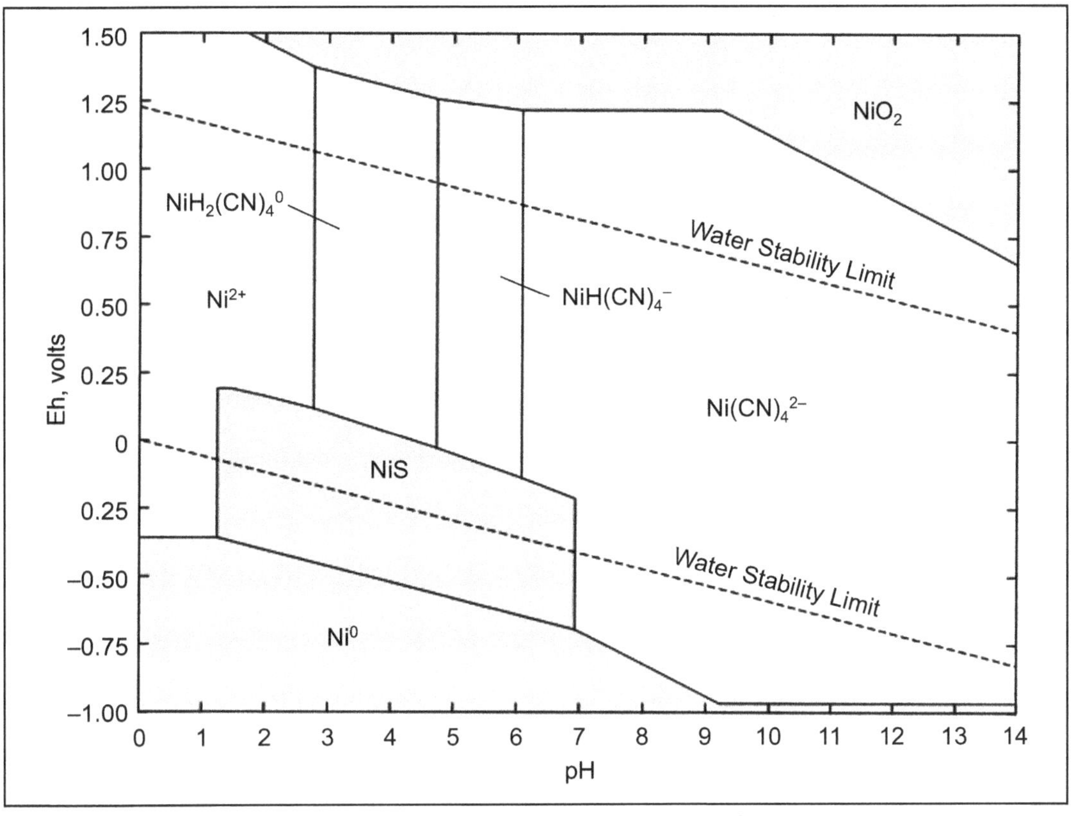

FIGURE 7.3 Eh-pH diagram for Ni-S-CN-H_2O system (a[Ni] $10^{-4.1}$, a[S] $10^{-4.1}$, a[CN] $10^{-2.4}$, 25°C)

predicted at low pH, but in practice, this requires elevated temperature or the presence of a strong competing ligand.

An Eh-pH diagram for the aqueous nickel-cyanide-sulfide system is shown in Figure 7.3. This diagram suggests the nickel-sulfides are soluble in cyanide solution at alkaline pH without the need for oxidizing conditions. At a pH less than 7.0, nickel-sulfide then becomes insoluble under reducing conditions. As noted earlier, free nickel Ni^{2+} is thermodynamically predicted at low pH, but this does not actually occur in practice because of kinetic limitations.

The Odor of Ammonia in Carbon Elution

Activated carbon that has been loaded with gold and silver is eluted at high temperature (140°C) and pressure (4–5 bar) with a solution of NaCN (0.1%) and NaOH (1%–2%). In the elution step, gold and silver are stripped from the carbon and then later recovered in crude bullion form after electrowinning or zinc cementation. At the elevated temperature in the elution step, cyanide hydrolyzes in the presence of activated carbon to yield formate ($HCOO^-$) and ammonia (NH_3).

$$CN^- + 2H_2O \rightarrow HCOO^- + NH_3$$

The rate of this reaction is temperature dependent. At 20°C, the half-life of cyanide is 3.5 years, but at temperatures of 100°C and 16v0°C, the half-life decreases to 8.1 hours and 7 minutes, respectively. This hydrolysis reaction is the cause of the familiar ammonia odor that is often present around carbon elution circuits.

Source: Marsden and House 2006; Sibrell et al. 1994

29
Cu
Copper
63.546

8 Copper

The most common oxidation state for copper is +2, and this is the state normally encountered in non-cyanide solutions. However, alkaline cyanide leach solutions are a bit unusual with respect to copper, in that copper is present only in the +1 oxidation state. Any Cu^{2+} entering into solution is rapidly reduced to Cu^{+}, which then forms soluble complexes with cyanide. The reaction of Cu^{2+} consumes cyanide by oxidizing it to cyanate according to the following reaction:

$$2Cu^{2+} + 7CN^{-} + 2OH^{-} \rightarrow 2Cu(CN)_3^{2-} + OCN^{-} + H_2O$$

Stability constants for the common copper-cyanide complexes are shown in Table 8.1.

TABLE 8.1 Stability constants for copper-cyanide complexes

Equilibrium Reaction	log β (25°C)	Reference
$Cu^{+} + 2CN^{-} \rightleftharpoons Cu(CN)_2^{-}$	23.94	Izatt et al. 1967
$Cu^{+} + 3CN^{-} \rightleftharpoons Cu(CN)_3^{2-}$	29.24	Izatt et al. 1967
$Cu^{+} + 4CN^{-} \rightleftharpoons Cu(CN)_4^{3-}$	30.74	Izatt et al. 1967
$Cu^{+} + CN^{-} \rightleftharpoons CuCN_{(s)}$	19.74	Lu et al. 2002

In any cyanide solution containing copper, the soluble complexes are some combination of $Cu(CN)_2^{-}$, $Cu(CN)_3^{2-}$, and $Cu(CN)_4^{3-}$. Higher molar ratios of cyanide to copper (CN:Cu) and higher pH levels favor the $Cu(CN)_3^{2-}$ and $Cu(CN)_4^{3-}$ complexes. Some level of free cyanide is always in equilibrium with these copper complexes, although the level can be quite low if the CN:Cu ratio is low.

In the paragraphs that follow, the molar ratio of CN:Cu is frequently referenced. When considering the cyanide concentration to be used in this ratio, it is important to recognize this is weak acid dissociable (WAD) cyanide and not cyanide determined from a conventional silver nitrate titration. Results from a conventional silver nitrate titration are not appropriate when determining the molar ratio of CN:Cu. A discussion in this regard is provided in Chapter 14.

Copper is increasingly problematic and expensive to manage in cyanide leaching operations. At an order-of-magnitude level, gold is often present in ores at about 1–10 g/t (grams per metric ton), while copper can be present at approximately 100–10,000 g/t. This means copper grades can be about two to three orders of magnitude greater than gold grades. But in terms of value, copper is worth about four orders of magnitude less than gold.* So there is a double disadvantage in that copper is often present at much higher grades than gold but is worth substantially less than gold.

Copper is seldom recovered as a salable product when gold and silver ores are being leached with cyanide, and so copper is often considered a nuisance metal. An important exception to this is in operations where copper is recovered by flotation ahead of a cyanide leach circuit. But even in this case, the residual copper present in flotation tailings is often challenging to manage in a cyanide leach circuit. Unfortunately, there are several copper minerals that are soluble in cyanide and are leached from ores along with gold and silver. Research and development work continues to be performed to address the high costs of dealing with copper at gold and silver leaching operations.

Elevated concentrations of copper in leach solution result from leaching ores containing copper minerals, such as those listed in Table 8.2. In actual practice, there is a high level of variability in the solubilities of the minerals listed in Table 8.2. Solubility depends on many factors, such as particle size distribution, mineral grain size, cyanide concentration, leach time, temperature, and oxygen availability. The general observation is that chalcopyrite and chrysocolla have lower solubilities, enargite and tetrahedrite have intermediate solubilities, while the remainder of the minerals have higher solubilities. Whenever an ore is being considered for treatment by cyanidation, the copper grades should be assayed, and mineralogical examinations should be performed to identify the specific minerals present. Assays for cyanide-soluble copper are also useful.

TABLE 8.2 Reported solubilities of copper minerals by cyanide

Mineral	Formula	Solubility, %	
		Hedley and Tabachnick 1958	Lower and Booth 1965
Azurite	$2CuCO_3{\cdot}Cu(OH)_2$	94.5–100	91.8
Bornite	$FeS{\cdot}2Cu_2S{\cdot}CuS$	70–100	96.0
Chalcocite	Cu_2S	90.2–100	54.4–92.6
Chalcopyrite	$CuFeS_2$	5.6–8.2	5.8
Chrysocolla	$CuSiO_3$	11.8–15.7	–
Covellite	CuS	–	53.2–95.6
Cuprite	Cu_2O	85.5–100	82.7–96.6
Enargite	$3CuS{\cdot}As_2S_5$	65.8–75.1	–
Malachite	$CuCO_3{\cdot}Cu(OH)_2$	90.2–100	76.9–99.7
Metallic copper	Cu	90–100	–
Tetrahedrite	$4Cu_2S{\cdot}Sb_2S_3$	21.9–43.7	–

An Eh-pH diagram for the aqueous copper-cyanide system is shown in Figure 8.1. Under the stated conditions, a large predominance region exists for the $Cu(CN)_3^{2-}$ complex in the Eh-pH range where gold and silver leaching are practiced. In this same region, the $Cu(CN)_2^-$ and

* Current commodity pricing (first quarter of 2023) for gold is about US$1,900 per ounce (equivalent to $27,700 per pound) and for copper about US$4.00 per pound (equivalent to $0.27 per ounce).

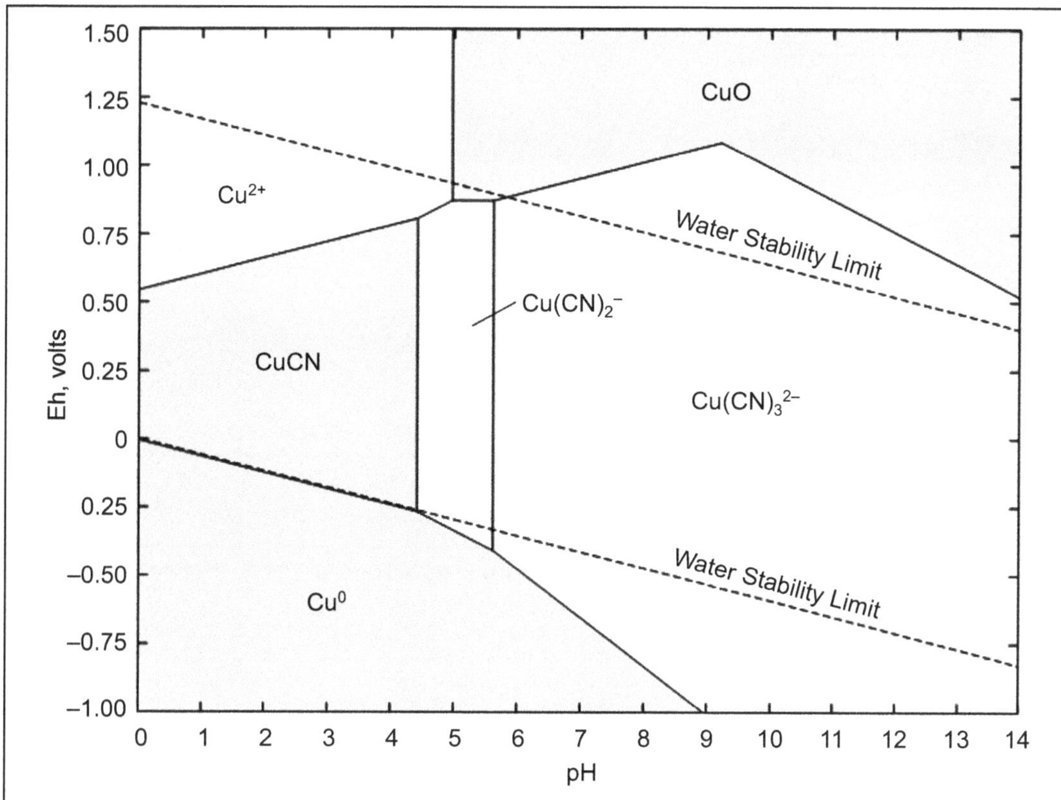

FIGURE 8.1 Eh-pH diagram for Cu-CN-H_2O system (a[Cu] $10^{-2.3}$, a[CN] $10^{-1.7}$, 25°C)

$Cu(CN)_4^{3-}$ complexes are also present but at non-predominant concentrations. A higher level of cyanide would be required for the $Cu(CN)_4^{3-}$ complex to become predominant at pH 10–11. The $Cu(CN)_2^-$ complex becomes predominant when the pH decreases below about 6. At a low pH, the solid $CuCN_{(s)}$ precipitates, but if thiocyanate is present in solution, the solid $CuSCN_{(s)}$ may form along with or in preference to $CuCN_{(s)}$.

A pH-distribution diagram for the aqueous copper-cyanide system is shown in Figure 8.2. The concentrations of 500 mg/L cyanide and 300 mg/L copper correspond to a molar ratio of about 4:1 for CN:Cu. At pH 10.5, there is a low concentration of $Cu(CN)_2^-$, only about 0.2 mg/L as Cu. There is also approximately 110 mg/L of free cyanide in solution, which is sufficient to ensure that gold and silver leaching and recovery operations are not starved for free cyanide. It is clear from Figure 8.2 that lower pH levels increase the proportion of $Cu(CN)_2^-$ in solution. Equilibrium between the various copper-cyanide complexes is established relatively quickly in leach solutions (i.e., kinetic reactions among the copper-cyanide complexes are relatively rapid).

The concentrations of free cyanide and the $Cu(CN)_2^-$ complex in leaching and recovery circuits are important to understand. At alkaline pH, if there is a lack of free cyanide in solution, then there are likely to be elevated concentrations of the $Cu(CN)_2^-$ complex and reduced concentrations of the $Cu(CN)_3^{2-}$ and $Cu(CN)_4^{3-}$ complexes. Elevated concentrations of $Cu(CN)_2^-$ are usually indicative of leach solutions deficient in free cyanide and thus inhibitory against effective gold and silver leaching. These solutions are often called *cyanide-starved.*

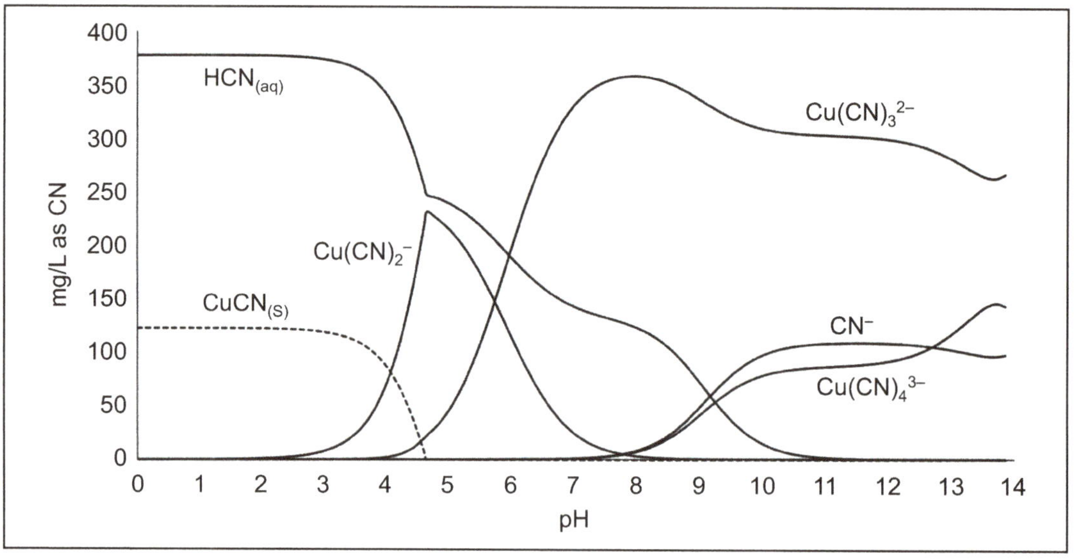

FIGURE 8.2 pH-distribution diagram for Cu-CN-H_2O system (500 mg/L CN^-, 300 mg/L Cu^+, 25°C) (CN:Cu molar ratio 4:1)

Sulfidization Process for Removing Copper

Removing copper from cyanide leach solution is possible through the dosing of sulfide at acidic pH. This process was described many years ago but has more recently come into practice as the sulfidization-acidification-recycling-thickening (SART) process. Copper is removed and recovered as a solid in the form of Cu_2S (synthetic chalcocite), as shown in the following reactions. The precipitated copper is filtered and then sold to a smelter. The remaining solution is neutralized with lime so it can safely be reused for leaching.

$$2Cu(CN)_3^{2-} + 2.5H_2SO_4 + NaHS \rightarrow Cu_2S_{(s)} + 2.5SO_4^{2-} + Na^+ + 6HCN_{(aq)}$$

$$6HCN_{(aq)} + 3Ca(OH)_2 \rightarrow 6CN^- + 3Ca^{2+} + 6H_2O$$

Zinc, silver, mercury, and cadmium are also removed and recovered in this manner. The economic benefits come from recovering copper as a salable product and from regenerating free cyanide. Cyanide that was previously bound to copper is now liberated as free cyanide and can be recycled to the leach circuit. This reduces the amount of new NaCN that must be purchased.

The process has been implemented at several cyanidation operations worldwide where copper in leach solution is problematic. Great caution must be exercised during design and operation because the cyanide solutions are acidified, and high concentrations of $HCN_{(aq)}$ are formed. The processes are completely enclosed to prevent the escape of $HCN_{(g)}$ and have been safely and successfully operated for many years.

Source: Leaver and Woolf 1929; Botz et al. 2011

In the presence of chalcopyrite and pyrite, a deficiency of free cyanide can even lead to these minerals functioning as preg-robbers (Rees and van Deventer 2000).

The $Cu(CN)_2^-$ complex is well known to have a high affinity for adsorption onto activated carbon and can heavily load onto carbon while interfering with the adsorption of gold and silver. This complex can also interfere with Merrill–Crowe circuits by consuming zinc while being reduced to metallic copper. It is not unusual for makeup NaCN to be dosed ahead of carbon adsorption and Merrill–Crowe circuits to shift the $Cu(CN)_2^-$ complex more toward the $Cu(CN)_3^{2-}$ and $Cu(CN)_4^{3-}$ complexes. When copper is loaded onto activated carbon, a cold sodium cyanide/sodium hydroxide (NaCN/NaOH) strip ahead of Au/Ag elution is often practiced to limit the copper content in final bullion.

When elevated levels of copper are present in a leach solution, Figure 8.3 can aid in evaluating the concentrations of WAD cyanide required to maintain various levels of free cyanide. As an example, if a leach solution has 500 mg/L of copper, and it is desired to maintain 50–100 mg/L of free cyanide for effective gold leaching, Figure 8.3 suggests the corresponding WAD cyanide concentration should be about 680–750 mg/L. Free cyanide levels of 100–200 mg/L may be desired for effective silver leaching, and in this case, the corresponding WAD cyanide concentration would increase to approximately 750–890 mg/L.

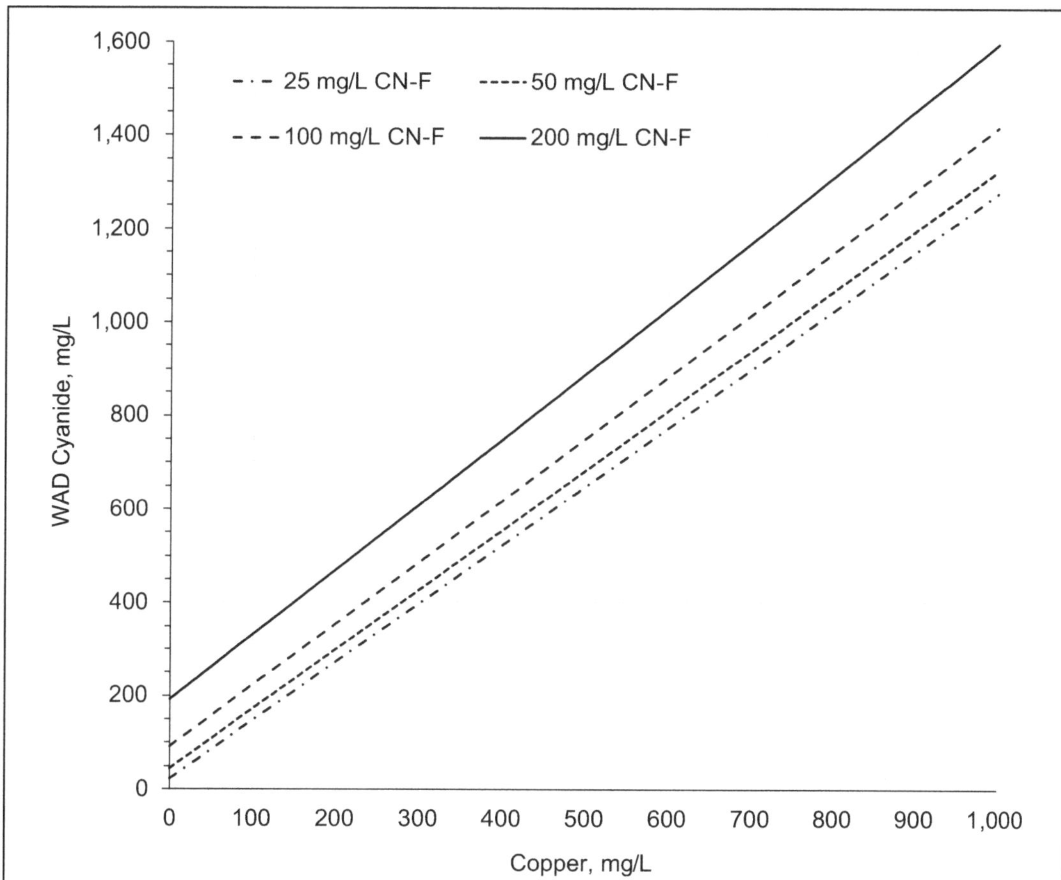

FIGURE 8.3 Concentrations of weak and dissociable (WAD) cyanide required to maintain various levels of free cyanide (CN-F) in solution (pH 11.0, 25°C)

Figure 8.4 is an alternative presentation of the data shown in Figure 8.3. In this figure, the molar ratio of CN:Cu is used rather than the WAD cyanide concentration on the vertical axis. Considering again 500 mg/L of copper in a leach solution, Figure 8.4 suggests the CN:Cu molar ratio should be about 3.3:1 to 3.7:1 to maintain 50–100 mg/L of free cyanide in solution. Doing the math, this equates to a WAD cyanide concentration of 680–760 mg/L, which is essentially the same as determined using Figure 8.3.

In practice, the target free-cyanide concentration or CN:Cu molar ratio should be adjusted, as appropriate, to balance the consumption of NaCN against the actual efficiencies observed for leaching and metal recovery. In the author's experience, a CN:Cu molar ratio in the range of 3.5:1 to 4.0:1 is generally suitable. This range usually provides sufficient free cyanide for leaching and minimizes the impact of copper in activated carbon and Merrill–Crowe circuits. A molar ratio of 4.0:1 has been recommended by others and observed to yield low loadings of copper onto activated carbon (Bailey 1987; Lower and Booth 1965). In cases where the copper concentrations are relatively low, the CN:Cu molar ratio has less utility and the use of Figure 8.3 is more appropriate.

With Figures 8.3 and 8.4, the presence of other metals in addition to copper has not been considered. Notably, this would be zinc, which can be present at elevated concentrations along with copper. If elevated concentrations of zinc are present, an additional portion of cyanide

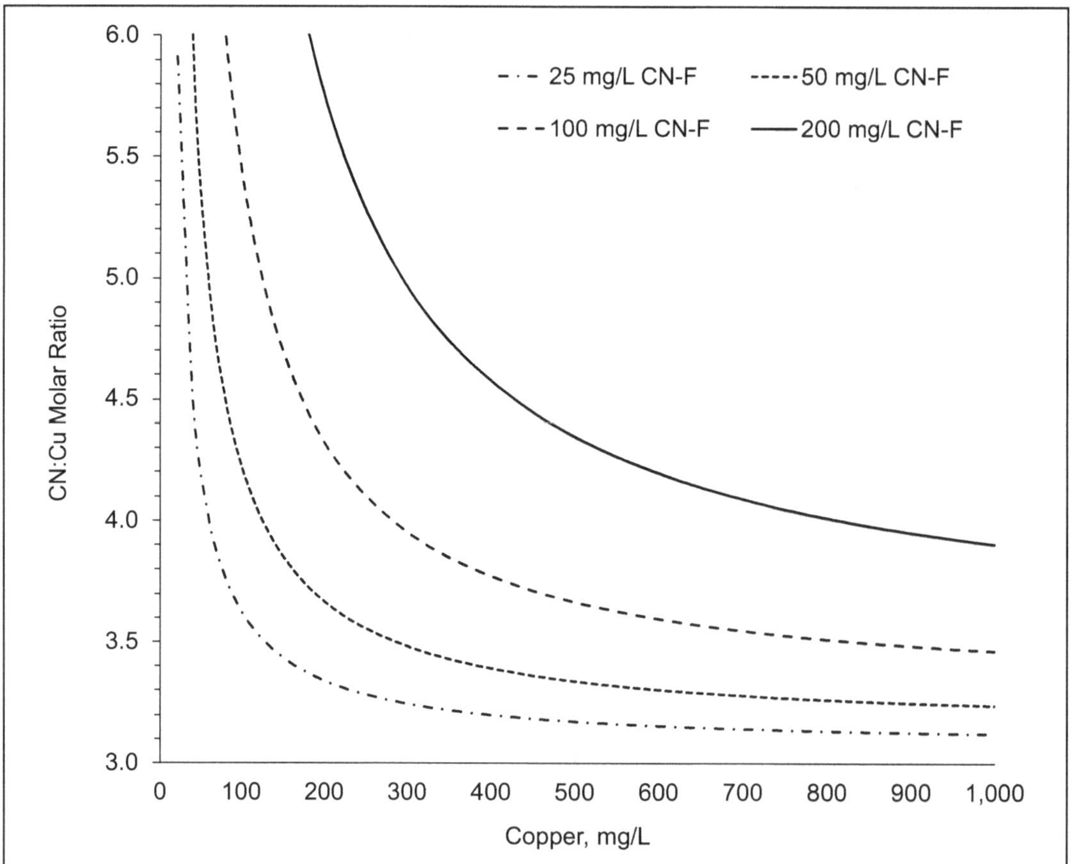

FIGURE 8.4 Molar ratios of CN:Cu required to maintain various levels of free cyanide (CN-F) in solution (pH 11, 25°C)

in solution is complexed with zinc and unavailable as free cyanide. As the zinc concentration increases, the concentration of $Cu(CN)_2^-$ also increases, and the concentration of free cyanide decreases. With leach solutions of this sort, it is helpful to calculate the detailed speciation using a software tool such as Geochemist's Workbench, STABCAL, Visual MINTEQ, or HSC Chemistry. An example of this type of calculation is given in Chapter 12.

An Eh-pH diagram for the aqueous copper-cyanide-sulfide system is shown in Figure 8.5. This diagram suggests copper-sulfides are soluble in cyanide solution at alkaline pH. At a pH of less than 7, copper-sulfides and copper-cyanide then become insoluble under reducing conditions. Relatively large regions of stability exist for Cu_2S, CuS, and CuCN solids at acidic pH and under reducing conditions. The sulfide solids actually present in this region likely consist of a range of different copper-sulfide minerals, as listed in Table 8.3.

Ores containing the mineral chalcopyrite ($CuFeS_2$) are widely encountered in gold and silver ore deposits, and it is interesting to observe the Eh-pH stability of chalcopyrite in a cyanide system. As shown in Figure 8.6, a large window of chalcopyrite stability develops when iron is added into the copper-cyanide system. This diagram suggests chalcopyrite is soluble under typical leaching conditions (aerated cyanide solutions at pH 10–11). But as noted previously, the solubility of chalcopyrite in actual practice is fairly limited because of slow kinetics (Table 8.2).

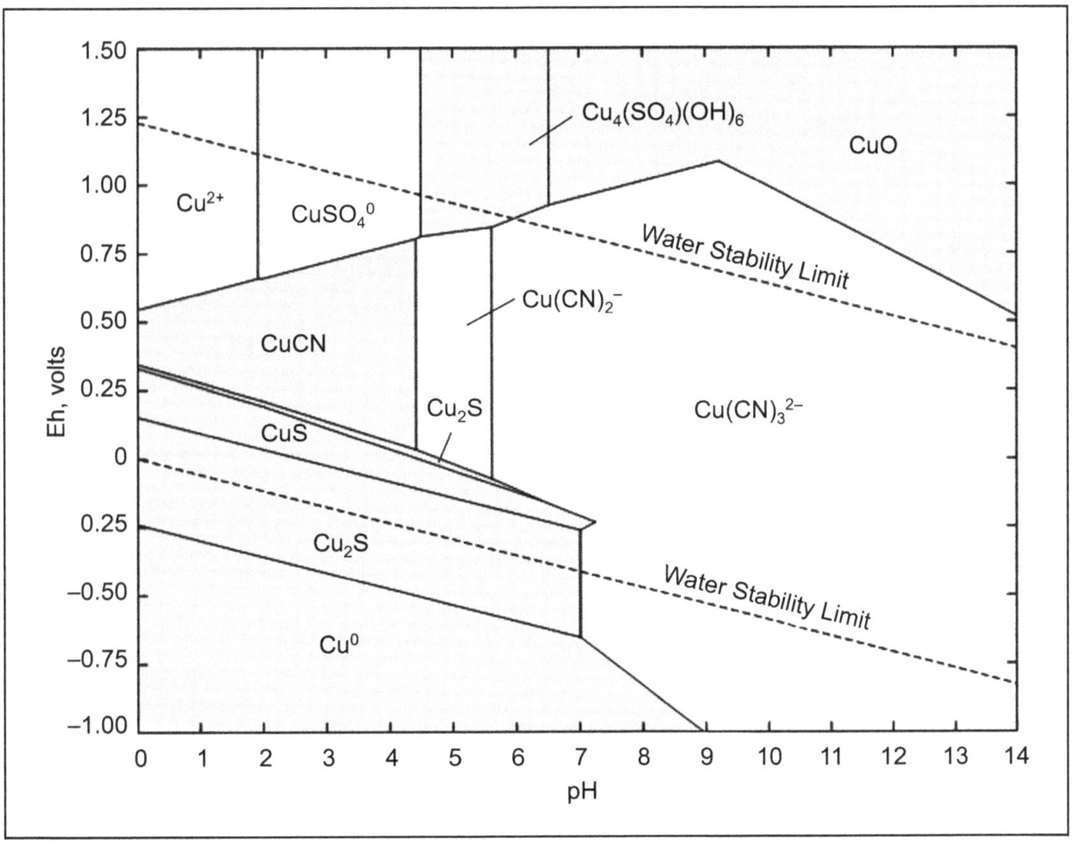

FIGURE 8.5 Eh-pH diagram for Cu-S-CN-H_2O system (a[Cu] $10^{-2.3}$, a[S] $10^{-2.3}$, a[CN] $10^{-1.7}$, 25°C)

TABLE 8.3 Composition of some copper-sulfide minerals

Mineral	Formula	Mole Ratio, Cu:S	Copper, wt %
Covellite	CuS	1	66.5
Yarrowite	Cu_9S_8	1.13	69
Spionkopite	$Cu_{39}S_{28}$	1.39	73.4
Anilite	Cu_7S_4	1.75	77.6
Digenite	Cu_9S_5	1.8	78.1
Djurleite	$Cu_{31}S_{16}$	1.94	79.3
Chalcocite	Cu_2S	2	79.9

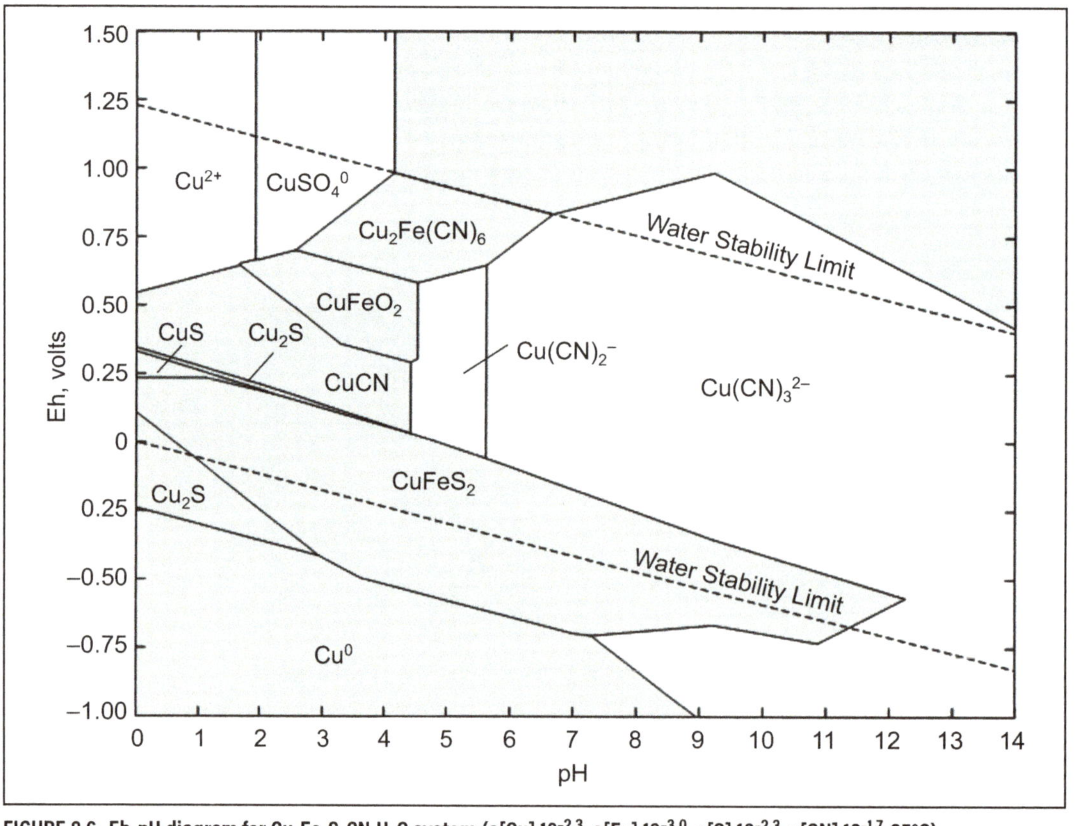

FIGURE 8.6 Eh-pH diagram for Cu-Fe-S-CN-H_2O system (a[Cu] $10^{-2.3}$, a[Fe] $10^{-3.0}$, a[S] $10^{-2.3}$, a[CN] $10^{-1.7}$, 25°C)

The solid $Cu_2Fe(CN)_6$ is shown in Figure 8.6 to exist under slightly acidic and oxidizing conditions. This solid likely exists in the hydrated form of $Cu_2Fe(CN)_6{\cdot}7H_2O$. Formation of this solid in some cyanide detoxification circuits is the mechanism by which iron-cyanides are removed (precipitated) from solution (Adams and Kyle 2000). The solid $[Cu(NH_3)_4][CuFe(CN)_6]{\cdot}H_2O$ may also form if ammonia is present.

Cyanide and Spies

When searching for books about cyanide, you will likely encounter the title *Between Silk and Cyanide: A Codemaker's War 1941–1945* by Leo Marks. This is not a chemistry or science book but is a historical account from World War II about secret codes and spies. The codes were printed onto silk, instead of paper. But silk can be concealed in clothing and does not make a rustling sound when a person is searched by the enemy. Unfortunately, such concealment was not always successful and that is when cyanide capsules were an unattractive option, but perhaps better than risking capture. It is an excellent book and worth your time.

30
Zn
Zinc
65.38

9 Zinc

The stable oxidation state for zinc is +2, and this is the state encountered in cyanide leach solutions. Zinc can be found at elevated concentrations in leach solution, either through the dissolution of zinc minerals, or more commonly, through zinc dust addition in Merrill–Crowe circuits. The chemistry associated with zinc dust addition (cementation) is discussed later in this chapter. Stability constants for the common zinc-cyanide complexes are shown in Table 9.1.

TABLE 9.1 Stability constants for zinc-cyanide complexes*

Equilibrium Reaction	log β (25°C)	Reference
$Zn^{2+} + CN^- \rightleftharpoons ZnCN^+$	5.34	NIST 2004†
$Zn^{2+} + 2CN^- \rightleftharpoons Zn(CN)_{2(aq)}$	11.07	Izatt et al. 1965
$Zn^{2+} + 3CN^- \rightleftharpoons Zn(CN)_3^-$	16.05	Izatt et al. 1965
$Zn^{2+} + 4CN^- \rightleftharpoons Zn(CN)_4^{2-}$	19.62	Izatt et al. 1965
$Zn^{2+} + 2CN^- \rightleftharpoons Zn(CN)_{2(s)}$	15.82	NIST 2004†

* Other values available for the stability constant (solubility product) of $Zn(CN)_{2(s)}$ include 11.03 calculated by the author using HSC Chemistry (Metso Outotec, n.d.); 12.59 in Sillén (1964); 15.5 in Smith and Martell (1976); and 15.59 in Knovel (2008) (25°C). The uncertainty in the value is noted in many publications.
† Calculated by the author using STABCAL (Montana Tech, n.d.) and NIST (2004).

The solubilities of various zinc minerals by cyanide are listed in Table 9.2. In actual practice, there is a relatively high level of variability that can be observed with the solubility of these minerals. Solubility depends on many factors, such as particle size distribution, mineral grain size, cyanide concentration, leach time, temperature, and oxygen availability. The general observation is that the minerals listed in Table 9.2 have somewhat low to intermediate levels of solubility.

An Eh-pH diagram for the aqueous zinc-cyanide system is shown in Figure 9.1. Of note is the very low Eh where elemental zinc metal Zn^0 is stable; it is this characteristic that makes zinc cementation effective for gold and silver recovery in a Merrill–Crowe circuit. The complex $Zn(CN)_4^{2-}$ normally predominates in leach solutions, being accompanied by lesser concentrations of the complex $Zn(CN)_3^-$. At neutral to slightly acidic pH, the solid $Zn(CN)_{2(s)}$ forms, while at a higher pH, the solid $ZnO_{(s)}$ may form depending on the concentrations of zinc and cyanide. The zinc-cyanides are relatively weak complexes and do not have nearly the impact on gold and silver leaching as does copper (van der Merwe and Breuer 2013).

TABLE 9.2 Solubility of zinc minerals by cyanide (45°C)

Mineral	Formula	Solubility, %
Willemite	Zn_2SiO_4	13.1
Calamine (hemimorphite)	$H_2Zn_2SiO_5$	13.4
Sphalerite	ZnS	18.4
Franklinite	$(Fe,Mn,Zn)O{\cdot}(Fe,Mn)_2O_3$	20.2
Hydrozincite	$3ZnCO_3{\cdot}2H_2O$	35.1
Zincite	ZnO	35.2
Smithsonite	$ZnCO_3$	40.2

Source: Leaver and Woolf 1929

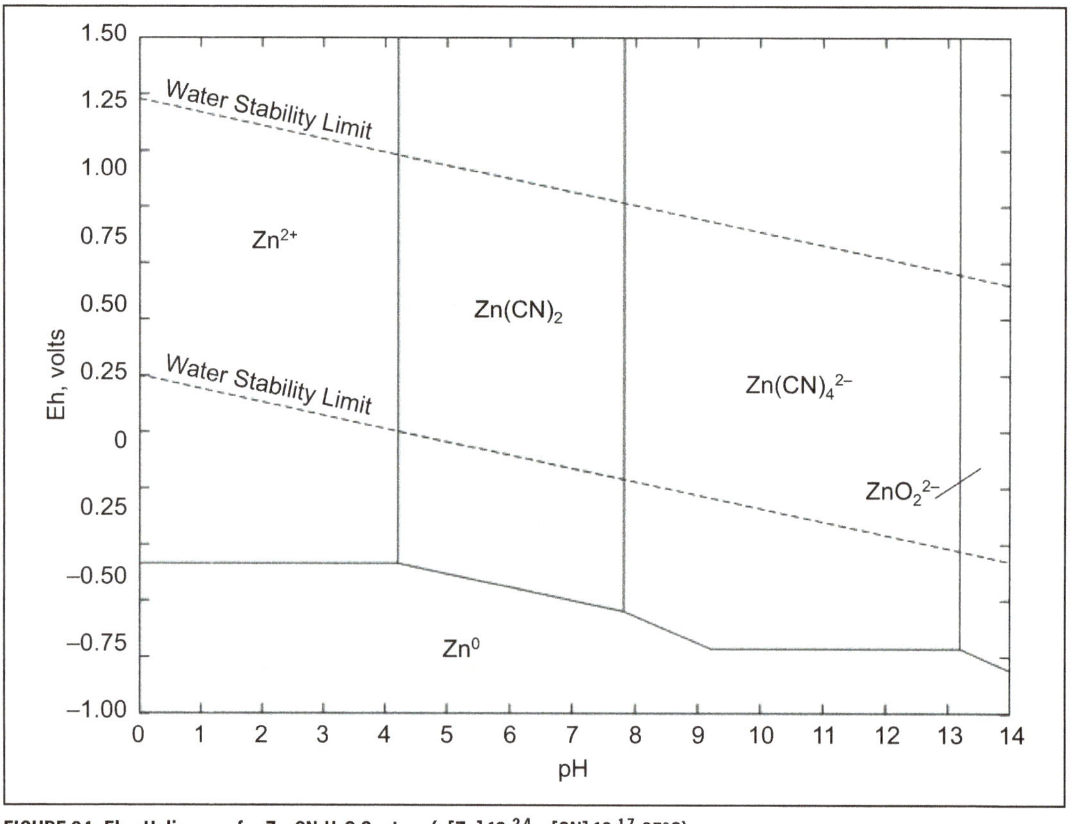

FIGURE 9.1 Eh-pH diagram for Zn-CN-H_2O System (a[Zn] $10^{-2.4}$, a[CN] $10^{-1.7}$, 25°C)

A pH-distribution diagram for the aqueous zinc-cyanide system is shown in Figure 9.2. The concentrations of 200 mg/L zinc and 400 mg/L cyanide correspond to a molar CN:Zn ratio of 5:1, which represents a reasonable excess of cyanide for maintaining zinc in solution. Zinc readily dissociates in acid solution to yield free cyanide and Zn^{2+}, while in highly alkaline solution, various soluble zinc-oxide/hydroxide complexes begin to form. There is a region in the near-neutral to acidic pH range where the solid $Zn(CN)_{2(s)}$ exists, and under conditions of higher pH, the solid $ZnO_{(s)}$ may form. The actual presence of any of these solids depends on the concentrations of zinc and cyanide and the solution pH. Equilibrium between the various

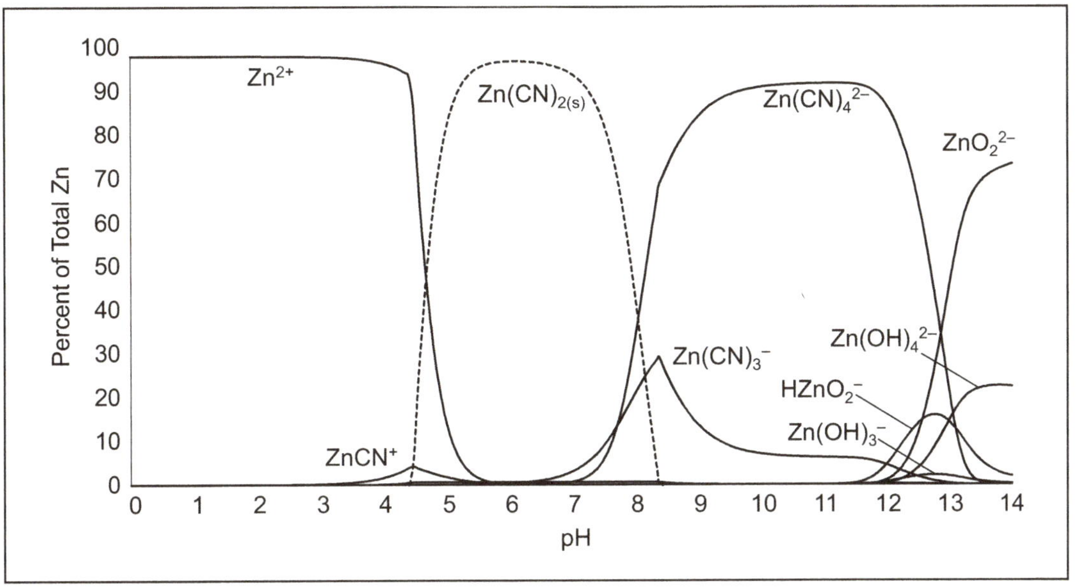

FIGURE 9.2 pH-distribution diagram for Zn-CN-H_2O system (Zn^{2+} 200 mg/L, CN^- 400 mg/L, 25°C)

History of the Merrill-Crowe Process

Zinc cementation was an early method used to recover gold and silver from leach solutions when cyanidation was first coming into use more than 125 years ago. The addition of elemental zinc to a pregnant solution creates a highly reducing environment (low Eh), in which gold and silver are reduced to their elemental states of Au^0 and Ag^0. As this occurs, the zinc is dissolved into solution as zinc-cyanide. The process was introduced in 1897 at Marysville, Montana in the plant of the Montana Mining Company and subsequently at the Homestake mine in South Dakota (United States).

This method of gold and silver recovery came to be known as the Merrill–Crowe process, named after two early innovators, C.W. Merrill and T.B. Crowe. Merrill substituted finely divided zinc dust for the larger zinc shavings that had previously been utilized, which significantly improved the process efficiency. The filters used to collect the precipitated metals are still sometimes called *Merrill presses*. Crowe implemented the use of deaeration towers to remove dissolved oxygen from pregnant solution ahead of zinc dust addition. This reduced the consumption of zinc through the side reaction with oxygen to form zinc oxide. The deaeration towers are still sometimes called *Crowe towers*.

The Merrill–Crowe process remains widely used today but primarily when silver concentrations are elevated in pregnant solution. With lower concentrations of silver, activated carbon adsorption is often the preferred method of recovery.

Source: Hampton 2002; MacFarren 1912; Clennell 1915

zinc complexes is established relatively quickly in leach solutions (i.e., kinetic reactions among the zinc-cyanide complexes are relatively rapid).

Solubility diagrams for the zinc-cyanide system are shown in Figure 9.3 for a zinc concentration of 200 mg/L and cyanide concentrations of 0–500 mg/L. Figure 9.3(a) shows the areas where solids in any amount are predicted to form, while Figure 9.3(b) shows the predominant species in the system as a function of pH and cyanide concentration.

In Figure 9.3(a), to prevent the formation of the solids $Zn(CN)_{2(s)}$ and $ZnO_{(s)}$ at pH 10–11, a minimum total cyanide concentration of about 350 mg/L is required. This corresponds to a molar CN:Zn ratio of 4.4:1. Avoidance of the insoluble regions for these solids is important for efficient operation of Merrill–Crowe circuits. Inadequate cyanide concentrations during cementation may be evidenced by reduced gold and silver recoveries, the formation of white-colored precipitates, and slowed filtration characteristics for the solids.

Zinc dust is used in Merrill–Crowe circuits to recover gold and silver in their elemental forms. The overall zinc cementation reactions are shown next. Cementation results in increased concentrations of the zinc-cyanide complexes in solution, shown as $Zn(CN)_4^{2-}$ in the reactions. Note that only a fraction of the zinc added in a Merrill–Crowe circuit actually reacts as shown; much of the added zinc is consumed through side reactions or remains unreacted because of surface blinding.

$$2Au(CN)_2^- + Zn^0 \rightarrow 2Au^0 + Zn(CN)_4^{2-}$$

$$2Ag(CN)_2^- + Zn^0 \rightarrow 2Ag^0 + Zn(CN)_4^{2-}$$

For elemental zinc to cause the precipitation (cementation) of a metal from solution, the low Eh imparted by the zinc must be below the potential needed to reduce the metal to its elemental state. The electrode potentials for several metal-cyanide half-cell reactions are shown in Table 9.3. These potentials indicate the general tendency of the metals to be reduced under low Eh conditions to their elemental states.

Referencing Table 9.3, reduction of the mercury complex $Hg(CN)_4^{2-}$ to form elemental mercury ($Hg^0_{(l)}$) thermodynamically occurs at −0.22 V under the stated conditions. This is 0.98 V more positive than the potential of −1.20 V that can be imparted by elemental zinc (i.e., the driving force for cementation in this case is −0.98 V). Thus, mercury should be recovered by zinc dust addition in a Merrill–Crowe circuit. In practice, that is exactly what is observed. A driving force of around −0.5 V is generally needed to effectively recover a metal by cementation. Thus, mercury-cyanide, silver-cyanide, and gold-cyanide should all be effectively cemented by zinc. These complexes have potentials furthest from that of Zn^0 (i.e., the driving force for cementation of these metals is greater than −0.5 V). These data also suggest iron in the form of $Fe(CN)_6^{4-}$ would not normally be cemented by zinc. By experience, it is known that the copper-cyanide complex $Cu(CN)_2^-$ can be cemented to some extent in a Merrill–Crowe circuit (Hsu and Tran 1996).

To prevent excessive recovery of copper during zinc cementation, it is important to minimize the concentration of $Cu(CN)_2^-$ (i.e., ensure that copper-cyanide is primarily present as $Cu(CN)_3^{2-}$ and $Cu(CN)_4^{3-}$). This is done by controlling the free cyanide concentration in the feed solution such that a minimal amount of $Cu(CN)_2^-$ is present. An excess of free cyanide also ensures that zinc is dissolved into solution primarily as $Zn(CN)_4^{2-}$, rather than

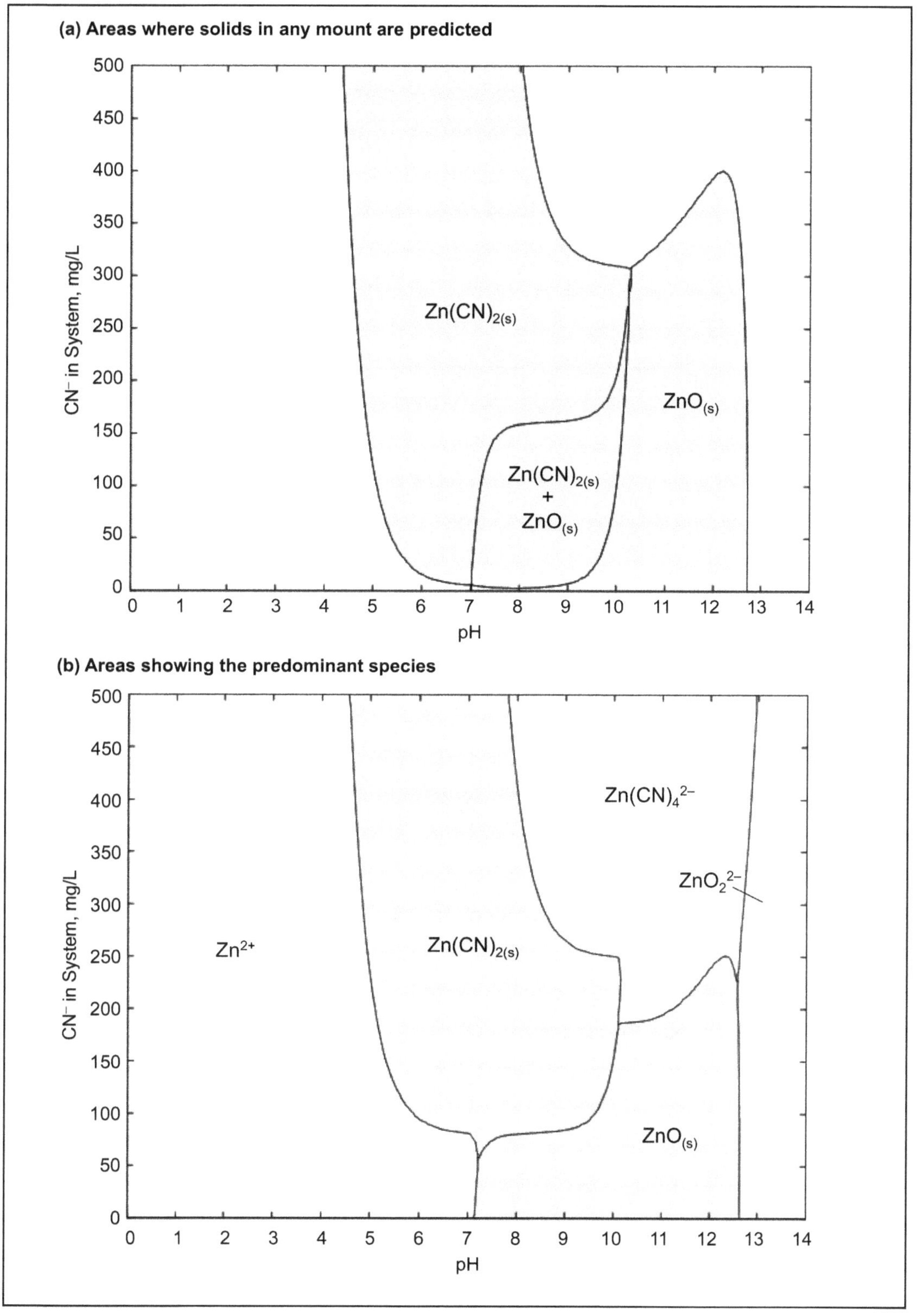

FIGURE 9.3 Solubility diagrams for Zn-CN-H_2O system (Zn^{2+} 200 mg/L, 25°C)

TABLE 9.3 Half-cell potentials for reduction of metal-cyanide complexes*

Reaction	a[Metal]	a[CN⁻]	E (V)
$O_{2(aq)} + 2H_2O + 4e^- \rightarrow 4OH^-$	—	—	+0.40
$Hg(CN)_4^{2-} \rightarrow Hg^0_{(l)}$	$10^{-4.6}$	$10^{-1.8}$	−0.22
$Ag(CN)_2^- \rightarrow Ag^0$	$10^{-3.7}$	$10^{-1.8}$	−0.42
$Au(CN)_2^- \rightarrow Au^0$	$10^{-4.6}$	$10^{-1.8}$	−0.67
$Co(CN)_5^{3-} \rightarrow Co^0$	$10^{-4.1}$	$10^{-1.8}$	−0.82
$Cu(CN)_2^- \rightarrow Cu^0$	$10^{-2.5}$	$10^{-1.8}$	−0.83
$2H_2O + 2e^- \rightarrow H_{2(g)} + 2OH^-$	—	—	−0.83
$Cd(CN)_4^{2-} \rightarrow Cd^0$	$10^{-4.4}$	$10^{-1.8}$	−0.85
$Cu(CN)_4^{3-} \rightarrow Cu^0$	$10^{-2.5}$	$10^{-1.8}$	−1.02
$Cu(CN)_3^{2-} \rightarrow Cu^0$	$10^{-2.5}$	$10^{-1.8}$	−1.04
$Ni(CN)_4^{2-} \rightarrow Ni^0$	$10^{-4.1}$	$10^{-1.8}$	−1.04
$Zn(CN)_4^{2-} \rightarrow Zn^0$	$10^{-2.5}$	$10^{-1.8}$	−1.20
$Fe(CN)_6^{4-} \rightarrow Fe^0$	$10^{-4.0}$	$10^{-1.8}$	−1.56

* The potentials shown in Table 9.3 were calculated by the author using Geochemist's Workbench with a custom thermodynamic database. Potentials are shown relative to the standard hydrogen electrode (SHE) at pH 10–11 and 25°C. The activity of cyanide corresponds to roughly 400 mg/L. The activities of the metals correspond to roughly 5 mg/L for Au, Cd, Co, Fe, Hg, and Ni; 20 mg/L for Ag; and 200 mg/L for Cu and Zn. Standard half-cell potentials are shown for the reduction of oxygen and water (Bard et al. 1985).

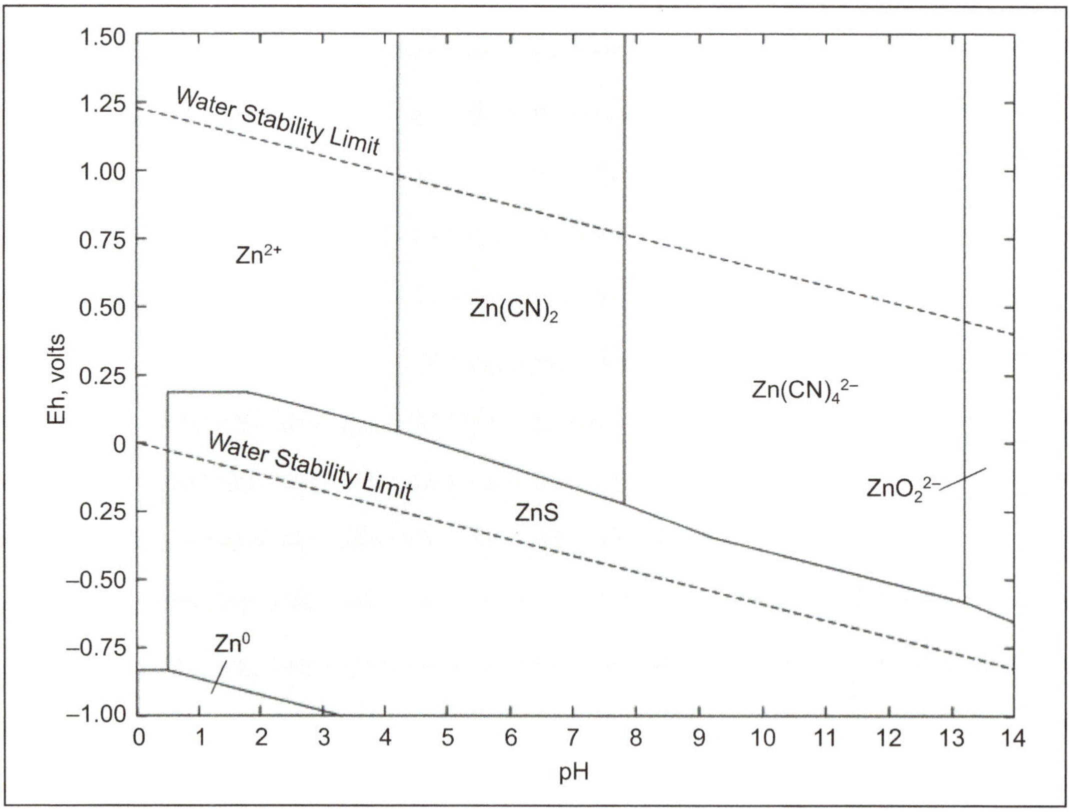

FIGURE 9.4 Eh-pH diagram for Zn-S-CN-H_2O system (a[Zn] $10^{-2.4}$, a[S] $10^{-2.4}$, a[CN] $10^{-1.7}$, 25°C)

$Zn(CN)_3^-$ or $Zn(CN)_2^0$, which provides the highest driving force for cementation. In this manner, it is possible to effectively operate Merrill–Crowe circuits with higher levels of copper.

An Eh-pH diagram for the aqueous zinc-cyanide-sulfide system is shown in Figure 9.4. There is a relatively large region of stability for ZnS under reducing conditions. Generally, ZnS (sphalerite) leaches slowly in alkaline cyanide solutions (Table 9.2) but with sufficient time, cyanide, and oxygen, leaching of sphalerite will occur.

48
Cd
Cadmium
112.414

10 Cadmium

The stable oxidation state for cadmium is +2, and this is the state encountered in cyanide leach solutions. Cadmium is generally found at low levels in cyanide leach solutions because it is usually only present at trace levels in gold and silver ores. This is primarily in association with sphalerite because cadmium can substitute for zinc in the sphalerite crystal lattice. Stability constants for the common cadmium-cyanide complexes are shown in Table 10.1.

TABLE 10.1 Stability constants for cadmium-cyanide complexes

Equilibrium Reaction	log β (25°C)	Reference
$Cd^{2+} + CN^- \rightleftharpoons CdCN^+$	6.01	Izatt et al. 1971
$Cd^{2+} + 2CN^- \rightleftharpoons Cd(CN)_{2(aq)}$	11.12	Izatt et al. 1971
$Cd^{2+} + 3CN^- \rightleftharpoons Cd(CN)_3^-$	15.65	Izatt et al. 1971
$Cd^{2+} + 4CN^- \rightleftharpoons Cd(CN)_4^{2-}$	17.92	Izatt et al. 1971
$Cd^{2+} + 2CN^- \rightleftharpoons Cd(CN)_{2(s)}$	10.3	Metso Outotec, n.d.*

* Calculated by the author using HSC Chemistry (Metso Outotec, n.d.).

An Eh-pH diagram for the aqueous cadmium-cyanide system is shown in Figure 10.1. In this diagram, the predominant form of cadmium at alkaline pH is shown to be $Cd(CN)_3^-$, although $Cd(CN)_2^0$ and $Cd(CN)_4^{2-}$ are present as well. The neutral complex $Cd(CN)_2^0$ is known to strongly adsorb onto activated carbon.

A pH-distribution diagram for the aqueous cadmium-cyanide system is shown in Figure 10.2. Cadmium readily dissociates in acid solution to yield free cyanide and Cd^{2+}, while in highly alkaline solution, the cadmium-hydroxide complexes begin to form, including the solid $Cd(OH)_{2(s)}$ at very high pH. Equilibrium with the cadmium complexes is established relatively quickly in leach solutions (i.e., kinetic reactions among the cadmium-cyanide complexes are relatively rapid).

An Eh-pH diagram for the aqueous cadmium-cyanide-sulfide system is shown in Figure 10.3. There is a relatively large region of stability for cadmium sulfide (CdS) solid under reducing conditions. Generally, CdS (greenockite) leaches slowly in alkaline cyanide solutions.

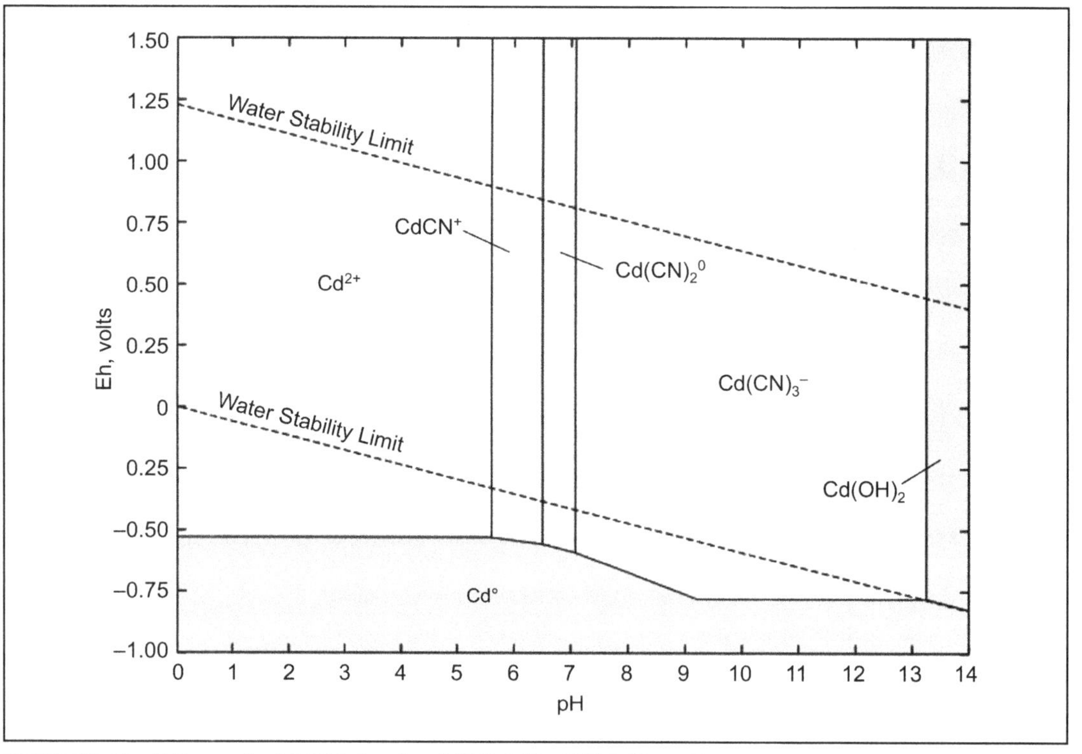

FIGURE 10.1 Eh-pH diagram for Cd-CN-H_2O system (a[Cd] $10^{-4.4}$, a[CN] $10^{-2.4}$, 25°C)

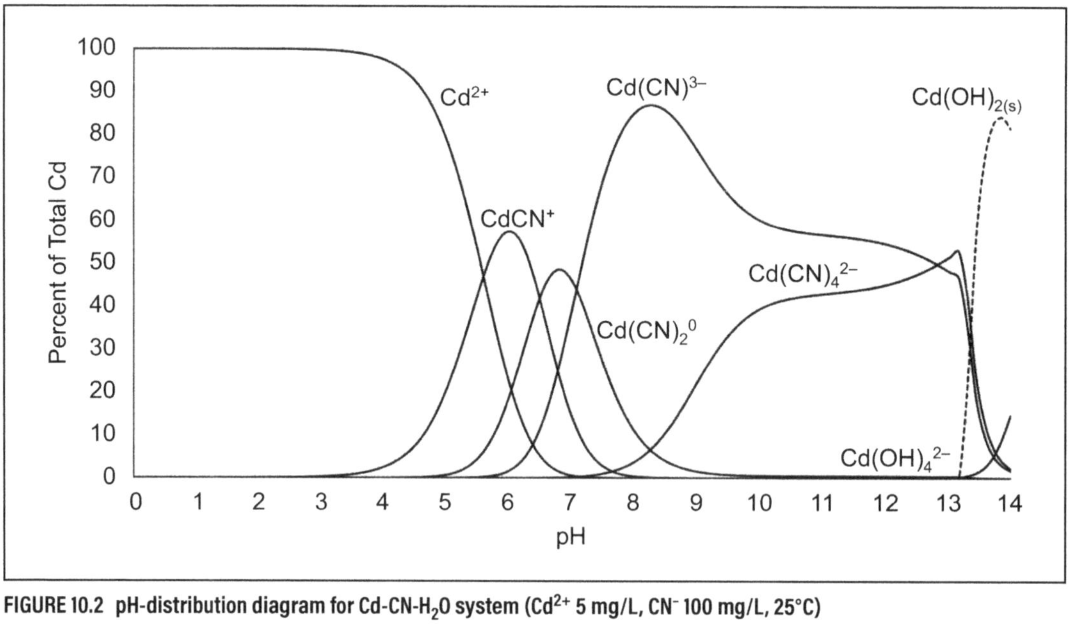

FIGURE 10.2 pH-distribution diagram for Cd-CN-H_2O system (Cd^{2+} 5 mg/L, CN^- 100 mg/L, 25°C)

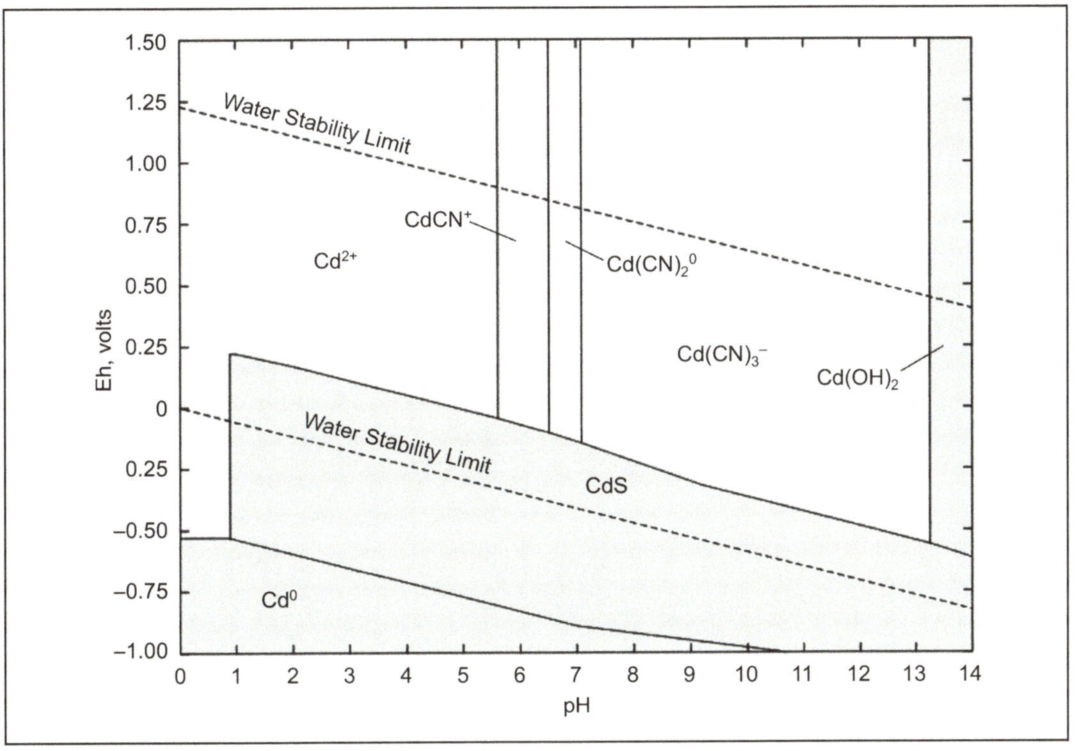

FIGURE 10.3 Eh-pH diagram for Cd-S-CN-H_2O system (a[Cd] $10^{-4.4}$, a[S] $10^{-4.4}$, a[CN] $10^{-2.4}$, 25°C)

More About Cobalt, Nickel, and Cadmium

Cobalt has its main uses in rechargeable battery electrodes and in making stainless steel alloys. The metallic cobalt alloys are stable at high temperature and are used to construct gas turbines and jet engines. Long ago, cobalt was used as an invisible ink. The ink consisted of cobalt chloride dissolved into water with some glycerin added, making a clear solution. The message written with this ink becomes visible upon heating and vaporization of the water and glycerin, thereby changing the color of cobalt to blue.

Nickel is found in large deposits around Sudbury, Ontario, Canada. The deposits are thought to have their origin in a large meteor impact. By far, the largest consumption of nickel comes with the making of corrosion-resistant stainless steels. The widely used stainless steels of Type 304 and Type 316 contain 8% and 11% nickel, respectively. The super alloys of Inconel 601, Incoloy 825, and Monel 400* contain about 58%–63% nickel, 38%–46% nickel, and 63% nickel plus cobalt, respectively.

Cadmium is a highly toxic metal, but it does have some important uses. Rechargeable NiCad batteries use nickel oxide and metallic cadmium as the electrodes. But they are being replaced by the less toxic nickel metal hydride batteries. Long ago, cadmium sulfide was used as a yellow pigment in paints and became one of Claude Monet's favored tones. Today, cadmium has a specialized use in the plating of fasteners for corrosion resistance, but this is limited to critical applications, such as with aircraft.

* Inconel®, Incoloy®, and Monel® are registered trademarks of Special Metals Corporation and its subsidiaries.
Source: Parsons and Dixon 2013; Gray 2009; Special Metals Corporation. n.d.

80
Hg
Mercury
200.59

Mercury

11

The common oxidation states for mercury are +1 and +2. The mercury +1 ion is unique and experimentally it has been found to exist as Hg_2^{2+} in solution. So, in equilibrium relationships the +1 oxidation state is often shown as Hg_2^{2+} rather than Hg^+. In oxidized and alkaline cyanide solutions, the soluble complexes all have the +2 oxidation state. Stability constants for the common mercury-cyanide complexes are shown in Table 11.1.

TABLE 11.1 Stability constants for mercury-cyanide complexes

Equilibrium Reaction	log β (25°C)	Reference
$Hg^{2+} + CN^- \rightleftharpoons HgCN^+$	17	Christensen et al. 1965
$Hg^{2+} + 2CN^- \rightleftharpoons Hg(CN)_2^0$	32.75	Christensen et al. 1965
$Hg^{2+} + 3CN^- \rightleftharpoons Hg(CN)_3^-$	36.11	Christensen et al. 1965
$Hg^{2+} + 4CN^- \rightleftharpoons Hg(CN)_4^{2-}$	38.97	Christensen et al. 1965
$Hg^{2+} + OH^- + CN^- \rightleftharpoons Hg(OH)(CN)_2^0$	29.71	Lingle 2000
$Hg^{2+} + 2CN^- \rightleftharpoons Hg(CN)_{2(s)}$	33.51	Flynn and McGill 1995
$Hg_2^{2+} + 2CN^- \rightleftharpoons Hg_2(CN)_{2(s)}$	39.3	EPA 2006*

* Calculated by the author using STABCAL (Montana Tech, n.d.) and MINTEQA2 (EPA 2006).

The most common mercury-containing mineral is cinnabar (HgS), but elemental mercury can also be found in ores. Mercury is leached by cyanide and once in solution may require special management through retorting or precipitation as an insoluble solid. Elevated levels of mercury in leach solution are not common, but even at low levels, mercury can have positive and negative impacts on leaching and recovery (Marsden and House 2006; Nicol et al. 1987). Mercury is readily adsorbed by activated carbon and effectively recovered during zinc cementation. Trace amounts of mercury may enhance gold and silver leaching and improve their recovery through zinc cementation.

An Eh-pH diagram for the aqueous mercury-cyanide system is shown in Figure 11.1. The complexes $Hg(CN)_2^0$, $Hg(CN)_3^-$, and $Hg(CN)_4^{-2}$ predominate in this case, depending on pH. The complexes $Hg(CN)_4^{2-}$ and $Hg(CN)_3^-$ readily dissociate at neutral to acidic pH to yield $Hg(CN)_2^0$. However, the $Hg(CN)_2^0$ complex is quite stable and resists further dissociation unless a ligand or elevated temperatures are employed. For the analysis of weak acid dissociable (WAD) cyanide using membrane separation techniques at room temperature, a

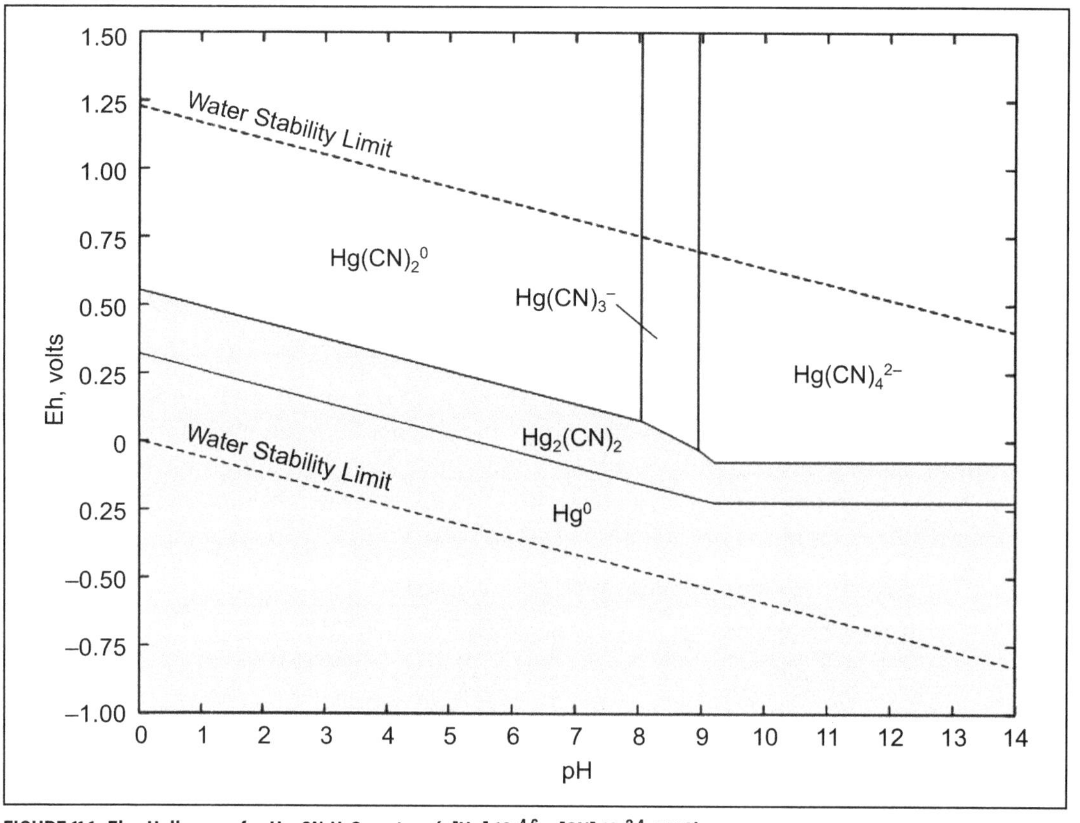

FIGURE 11.1 Eh-pH diagram for Hg-CN-H_2O system (a[Hg] $10^{-4.6}$, a[CN] $10^{-2.4}$, 25°C)

sulfur-based ligand is added as the competing ligand to complex with mercury and release free cyanide. Under reducing conditions, there is a wide region of stability for elemental Hg^0 liquid and the solid $Hg_2(CN)_{2(s)}$.

A pH-distribution diagram for the aqueous mercury-cyanide system is shown in Figure 11.2. In the pH range of 10–11, which is commonly utilized for cyanide leaching, the complexes $Hg(CN)_4^{2-}$ and $Hg(CN)_3^-$ are primarily present, along with a lesser amount of $Hg(CN)_2^0$. The complex $Hg(CN)_2^0$ is noteworthy in that it is readily adsorbed by activated carbon. As evidenced in Figure 11.2, even at a strongly acidic pH, the $Hg(CN)_2^0$ complex remains stable. The complex $HgCN^+$ is present in solution only at low concentrations.

An Eh-pH diagram for the aqueous mercury-cyanide-sulfide system is shown in Figure 11.3. There is a region of stability that exists for HgS solid under reducing conditions, with most of the region being present at acidic pH conditions. Under more alkaline pH conditions, the stability region for HgS narrows, and there is only a small region present at pH 10–11. Under reducing conditions at pH 10–11, the solids $Hg_2(CN)_2$ and HgS exist along with elemental Hg^0 liquid. Generally, HgS (cinnabar) and elemental Hg^0 liquid (quicksilver) leach fairly slowly in cyanide solutions, and their leaching consumes dissolved oxygen.

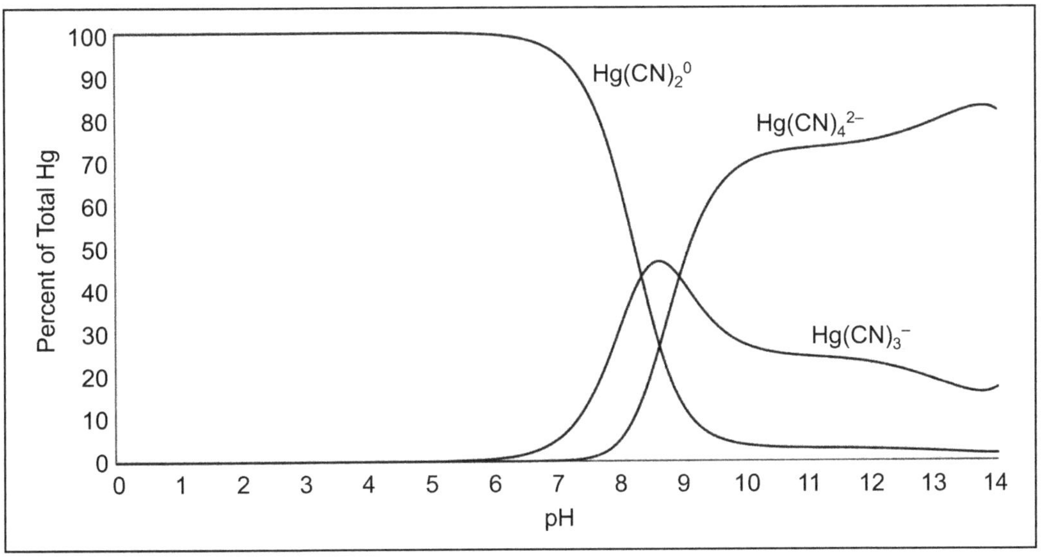

FIGURE 11.2 pH-distribution diagram for Hg-CN-H_2O system (Hg^{2+} 5 mg/L, CN^- 100 mg/L, 25°C)

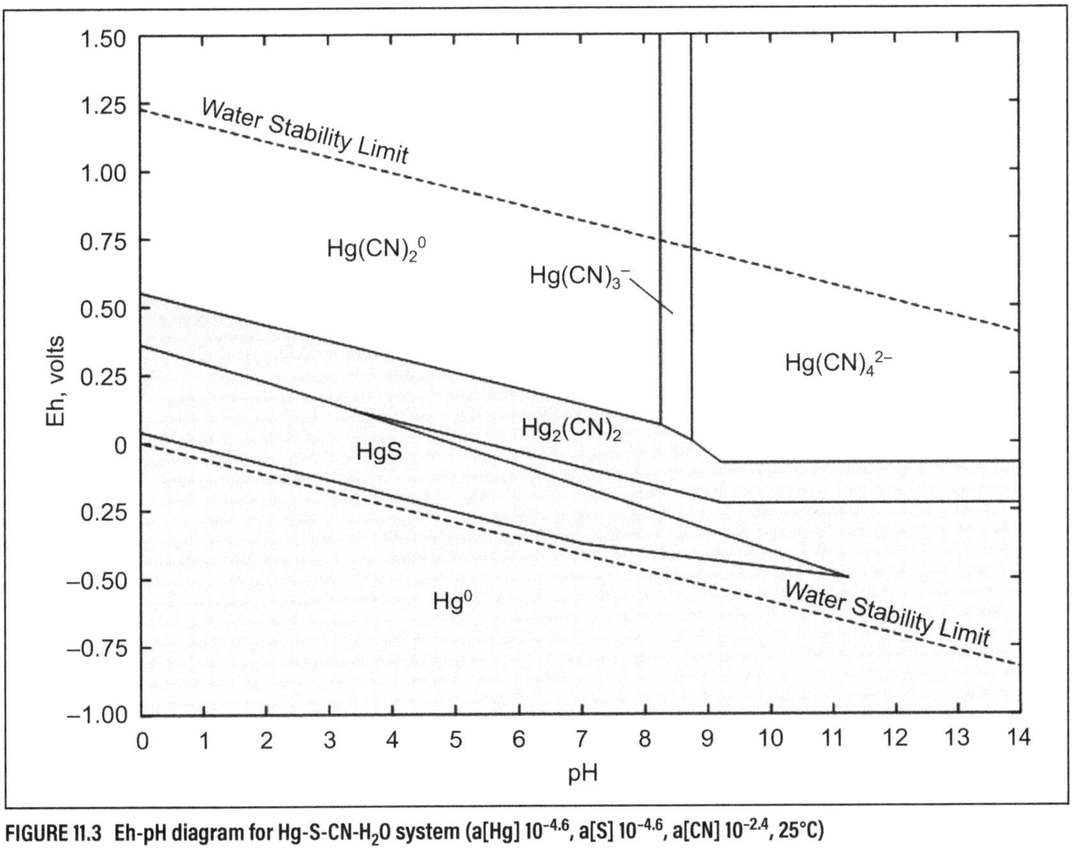

FIGURE 11.3 Eh-pH diagram for Hg-S-CN-H_2O system (a[Hg] $10^{-4.6}$, a[S] $10^{-4.6}$, a[CN] $10^{-2.4}$, 25°C)

Mixed Complexes

12

Parameter	Value, mg/L
Cyanide, total	800
Gold	2
Silver	5
Iron	10
Cobalt	0.5
Nickel	1
Copper	400
Zinc	150
Cadmium	0.2
Mercury	0.1

TABLE 12.1 Composition of example pregnant leach solution at pH 10.5

In this chapter, the metal-cyanide chemistry discussed in the previous chapters is brought together by considering a hypothetical (but realistic) cyanide leach solution. The chemistry shown in Table 12.1 could reflect a pregnant solution arising from leaching of a gold ore containing silver and a notable amount of cyanide-soluble copper. The elevated zinc concentration occurs because of a Merrill–Crowe zinc cementation circuit being used for gold and silver recovery. This leach solution has an elevated concentration of total cyanide and low levels of other metals.

The detailed speciations of the cyanide complexes in this solution were calculated and the results are summarized in Table 12.2. As indicated, the majority of cyanide in this solution is complexed with copper (63% of the total) and zinc (29% of the total). This is expected based on the composition shown in Table 12.1.

Table 12.2 shows cyanide is distributed through 24 complexes, 6 of which account for more than 95% of the cyanide in solution. This type of calculation has been used to evaluate the various routes of cyanide consumption in leach circuits (Adams 2001).

The calculated free cyanide concentration of 34 mg/L in this solution might be marginal in regard to ensuring that the leach circuit is not starved for free cyanide, and thus inhibiting gold and silver leaching. The zinc-cyanide concentration is relatively high (230 mg/L as CN), and this provides some cyanide that is available and effective toward gold and silver leaching (van der Merwe and Breuer 2013). Plant trials or bottle roll testing in this situation is helpful to determine if a slightly higher cyanide concentration might be beneficial for gold and silver leaching.

The concentration of $Cu(CN)_2^-$ in the solution is about 1.1 mg/L (as Cu). This is a relatively low concentration and not likely to overly interfere with the Merrill–Crowe zinc cementation step. However, if activated carbon were being utilized, there could be a notable loading of copper onto the carbon. In this case, there may be benefit in dosing makeup sodium cyanide (NaCN) ahead of carbon adsorption to limit copper loading.

The difference between the total cyanide and weak acid dissociable (WAD) cyanide concentrations listed in Table 12.2 is about 30 mg/L. This difference is all attributable to the cyanide

TABLE 12.2 Detailed speciation for example pregnant leach solution (25°C)*

Species	Cyanide mg/L	Cyanide % total	Metal mg/L	Metal % total
Free cyanide	**34.08**	**4.26**	—	—
CN^-	32.61	4.08		
$HCN_{(aq)}$	1.47	0.18		
Gold	**0.53**	**0.07**	**2**	**100**
$Au(CN)_2^-$	0.53	0.07	2	100
Silver	**2.48**	**0.31**	**5**	**100**
$Ag(CN)_2^-$	2.28	0.28	4.72	94.45
$Ag(CN)_3^{2-}$	0.2	0.03	0.28	5.55
Iron	**27.95**	**3.49**	**10**	**100**
$Fe(CN)_6^{4-}$	0	0	0	0
$Fe(CN)_6^{3-}$	27.95	3.49	10	100
Cobalt	**1.32**	**0.17**	**0.5**	**100**
$Co(CN)_6^{3-}$	1.32	0.17	0.5	100
Nickel	**1.77**	**0.22**	**1**	**100**
$Ni(CN)_4^{2-}$	1.77	0.22	1	100
Copper	**501.39**	**62.67**	**400**	**100**
$Cu(CN)_2^-$	0.92	0.12	1.13	0.28
$Cu(CN)_3^{2-}$	458.34	57.29	373.15	93.29
$Cu(CN)_4^{3-}$	42.12	5.27	25.72	6.43
Zinc	**230.28**	**28.79**	**150**	**100**
$ZnCN^+$	<0.01	<0.01	<0.01	<0.01
$Zn(CN)_2^0$	0.14	0.02	0.17	0.12
$Zn(CN)_3^-$	25	3.12	20.94	13.96
$Zn(CN)_4^{2-}$	205.15	25.64	128.88	85.91
Cadmium	**0.15**	**0.02**	**0.2**	**100**
$CdCN^+$	<0.01	<0.01	<0.01	0.01
$Cd(CN)_2^0$	<0.01	<0.01	<0.01	1.75
$Cd(CN)_3^-$	0.1	0.01	0.15	75.07
$Cd(CN)_4^{2-}$	0.04	0.01	0.05	23.16
Mercury	**0.04**	**0.01**	**0.1**	**100**
$HgCN^+$	<0.01	<0.01	<0.01	<0.01
$Hg(CN)_2^0$	<0.01	<0.01	0.01	13.56
$Hg(CN)_3^-$	0.02	<0.01	0.04	39.29
$Hg(CN)_4^{2-}$	0.02	<0.01	0.05	47.15
Total cyanide	**800**	**100**	—	—
WAD cyanide	**770.19**	**96.27**		
Free cyanide	**34.08**	**4.26**		

* Speciation was calculated by the author using Geochemist's Workbench with stability constants as listed in Chapters 2 through 11 and pO_2 0.2 atm. Generally similar results are obtained using other software tools.

bound to Au, Co, and Fe. These are the metals that form very stable complexes with cyanide and are only measured (or partially measured) using a total cyanide analytical technique (see the "Classifications of Cyanide" and "WAD and Total Cyanide" sections in Chapter 14).

The author estimates if this solution were titrated with silver nitrate using rhodanine color indicator that the titratable cyanide concentration would be around 370 mg/L as CN (for a related discussion, see the "Silver Nitrate Titration" section in Chapter 14). If the titration

Mercury Retorting

Mercury can be recovered in Merrill–Crowe and activated carbon adsorption circuits. If the amount of recovered mercury is high, it can be removed through a retorting process before the final metal product is sold as crude bullion. Elemental mercury is a liquid at room temperature (freezing temperature −39°C, boiling temperature 357°C), and it has a high vapor pressure at elevated temperatures, as shown in the following figure.

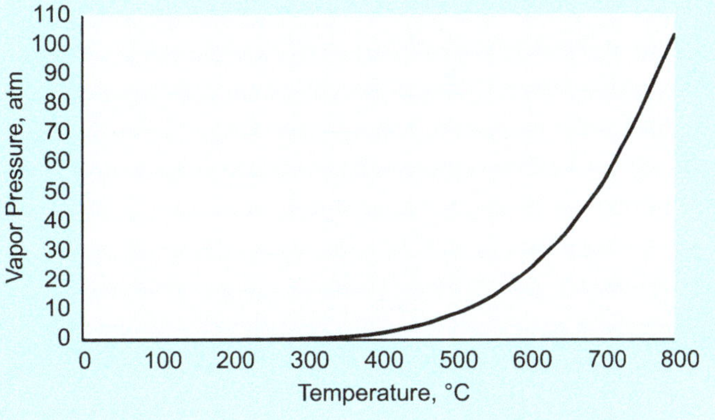

Data from Lange Dean

Vapor pressure of elemental mercury

During retorting, the precipitates or sludges containing mercury are heated to 600°–700°C to drive off mercury as a vapor. Mercury vapor is quite hazardous, so this is performed in a completely closed system under vacuum. The mercury vapor is then condensed to a liquid in a water-cooled condenser and collected into storage flasks. The retorting process yields final bullion with more than 99% of the mercury removed.

Mercury is unique in that it has a high vapor pressure for a metal and is a liquid at room temperature. Gallium is close, and with a freezing point of just 30°C, it will melt in your hand. At one time, mercury had been widely used in thermometers because of its low freezing temperature and high thermal expansion. But its toxicity led to the use of alcohol in thermometers, and more recently, to digital thermometers.

Source: Marsden and House 2006; Lide 2005; Dean 1992

result were interpreted as free cyanide rather than titratable cyanide, this could give the false impression that this solution contains a large excess of free cyanide. This is a cautionary example of why the silver nitrate titration method using a rhodanine end point can significantly overestimate the concentration of free cyanide when elevated levels of copper or zinc are present in solution. Only by calculating the detailed speciation of this solution can the true free cyanide concentration be reliably determined.

16
S
Sulfur
32.065

Sulfur Reactions

13

The chemistry of sulfur in cyanide leach circuits can be complex, perhaps even more complex than that of cyanide, and this chemistry becomes particularly important when sulfidic ores or concentrates are being leached. This chapter provides a brief introduction to the effects sulfur can have on leaching chemistry, with the aim of explaining why high consumptions of cyanide and oxygen can take place when sulfidic materials are being leached.

Sulfur is most often associated with sulfide-bearing minerals, which can have a wide range of reactivity toward cyanide. This reactivity can lead to the presence of many different solid and solubilized forms of sulfur, often referred to as *reduced sulfur species.* Reduced sulfur species are any sulfur compounds having an oxidation state (or valence) below that of sulfate. Sulfate is the most oxidized and thermodynamically stable form of sulfur in an aerated solution, having an oxidation state of +6. The most reduced form of sulfur is sulfide with an oxidation state of −2. Several of the more common reduced sulfur species with oxidation states between −2 and +6 are listed in Table 13.1.

The average oxidation states of sulfur are shown in Table 13.1 for each of the compounds. This is because there is uncertainty regarding the oxidation states of the individual sulfur atoms in some of the compounds. And in several of the compounds, sulfur is present in more than one oxidation state. For example, thiosulfate ($S_2O_3^{2-}$) is believed to have one sulfur atom in

TABLE 13.1 Sulfate and the common reduced sulfur species

Compound	Formula	Average S Oxidation State
Sulfide	S^{2-}	−2
Thiocyanate	SCN^-	−2
Polysulfides*	S_nS^{2-}	$-2 \div (n+1)$
Sulfur†	S_8	0
Thiosulfate	$S_2O_3^{2-}$	+2
Polythionates‡	$S_nO_6^{2-}$	$10 \div n$
Sulfite	SO_3^{2-}	+4
Sulfate	SO_4^{2-}	+6

* The value of *n* in the polysulfides generally ranges from 1 to 5, so that the average oxidation state of S ranges from −1 to −0.33.

† The stable form of elemental sulfur is generally S8 in an octamer ring.

‡ The value of *n* in the polythionates generally ranges from 3 to 7, so that the average oxidation state of S ranges from +1.43 to +3.33.

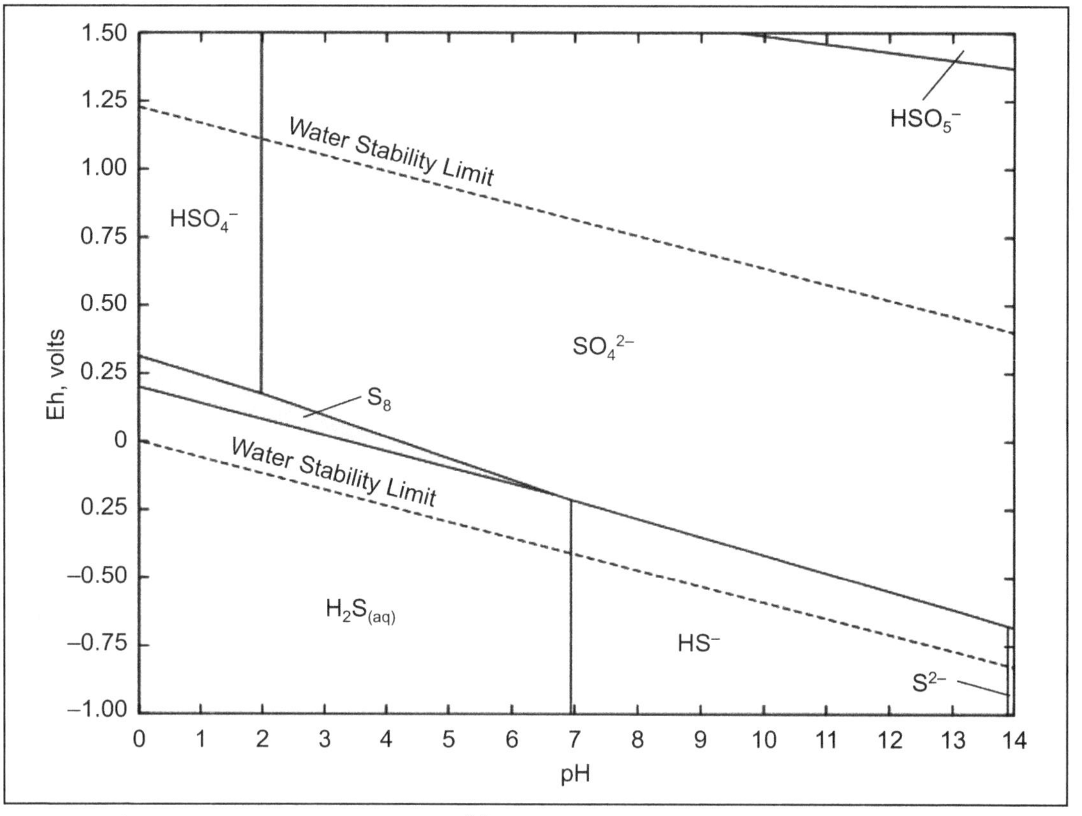

FIGURE 13.1 Eh-pH diagram for S-H_2O system (a[S] $10^{-2.0}$, 25°C)

the 0 oxidation state and one sulfur atom in the +4 oxidation state, thus yielding an average oxidation state of +2 for the two sulfur atoms (Suzuki 1999).

An Eh-pH diagram for the aqueous sulfur system is shown in Figure 13.1. Sulfide is predominant under reducing conditions as aqueous hydrogen sulfide ($H_2S_{(aq)}$), the bisulfide anion (HS^-), and the sulfide anion (S^{2-}). When sulfide is oxidized, several of the reduced sulfur species listed in Table 13.1 can form as intermediates. With continued oxidation, the most stable compound of sulfate will eventually be formed. The kinetics of oxidation are not necessarily rapid; therefore, it is common for a number of reduced sulfur species to be present in leach solutions exposed to sulfide minerals.

The key issues that arise when sulfide minerals are being leached include the following:

- Sulfide oxidizes to yield one or more of the reduced sulfur species listed in Table 13.1, thereby consuming oxygen. In some cases, oxygen consumption can become quite significant. Oxygen is necessary for gold and silver leaching and any side reactions that act to consume oxygen may inhibit gold and silver leaching.
- The reduced sulfur species may react with cyanide to form thiocyanate, which is an important pathway for the irreversible loss of cyanide. The term *irreversible* is used because thiocyanate does not revert back to cyanide, except under strongly oxidizing and acidic conditions (Botz et al. 2001).

- Metals associated with the sulfide minerals may react with cyanide to form various soluble and insoluble metal-cyanide complexes, as discussed in the previous chapters. Formation of some of these metal-cyanide complexes can also lead to an irreversible loss of usable cyanide. The best-known example is the formation of the iron-cyanides $Fe(CN)_6^{3-}$ and $Fe(CN)_6^{4-}$ caused by the reactions of cyanide with iron-sulfides.

The reaction of cyanide directly with sulfide is relatively slow. Only after sulfide is partially oxidized to another of the reduced sulfur species can the reaction rate with cyanide become appreciable. In general, the dominant reduced sulfur species resulting from sulfide oxidation are the polysulfides (primarily tetrasulfide and pentasulfide), thiosulfate, and the polythionates (primarily trithionate and tetrathionate) (Wilmot 1997; Wasserlauf and Dutrizac 1982). Trithionate and tetrathionate are the more stable of the polythionates under typical leach conditions; the higher polythionates undergo decomposition to form elemental sulfur, sulfate, thiosulfate, and/or a lower polythionate (Wasserlauf and Dutrizac 1982).

An increasing amount of oxygen is required as oxidation progresses from top to bottom in Table 13.1 (i.e., from sulfide toward sulfate). As an example, consider the oxidation of sulfide to form polysulfide with $n = 3$, as shown in the following reaction. For each 1 g of sulfide oxidized to tetrasulfide, there is a corresponding consumption of 0.37 g of oxygen.

$$4S^{2-} + 1.5O_2 + 6H^+ \rightarrow S_3S^{2-} + 3H_2O$$

Similarly, the oxidation of sulfide to form thiosulfate is shown in the next reaction:

$$S^{2-} + O_2 + H^+ \rightarrow \tfrac{1}{2}S_2O_3^{2-} + \tfrac{1}{2}H_2O$$

For each 1 g of sulfide oxidized to thiosulfate, there is a corresponding consumption of 1 g of oxygen. The complete oxidation of sulfide to sulfate requires double the amount of oxygen as compared to the oxidation of sulfide to thiosulfate. Eventually, all reduced sulfur species will be oxidized to sulfate if sufficient time and oxygen are available. But this does not necessarily occur in a cyanidation circuit because of limited reaction time or insufficient oxygen supply.

About the Sulfite in Wine

You may have noticed the warning label on your bottle of wine stating the product "Contains Sulfites." This warning is required if sulfur dioxide or sulfite is present in the wine at 10 ppm or greater. Why would sulfite be added to your wine? Sulfite is a reducing agent (see the "Eh-pH Diagrams" section in Chapter 1) and reacts to remove dissolved oxygen from wine. This extends the life of wine by preventing oxidization and the growth of unwanted bacteria. Sulfite is also used for this purpose in a variety of foods, including dried fruits and vegetables, cured meats, and jams and jellies. Sulfite is an effective preservative, but some people are sensitive to it, hence the reason for the warning.

Source: TTB 2023

Elemental sulfur, the polysulfides, thiosulfate, and the polythionates all react with cyanide to some extent to form thiocyanate according to the following reactions. The polysulfide pathway to thiocyanate likely dominates, as this reaction is relatively rapid at alkaline pH (Wilmot 1997; Luthy and Bruce 1979). When the reduced sulfur species react with cyanide in this manner, they function as cyanicides because they irreversibly consume cyanide.

$$S_8 + 8CN^- \rightarrow 8SCN^-$$

$$S_nS^{2-} + CN^- \rightarrow S_{n-1}S^{2-} + SCN^-$$

$$S_2O_3^{2-} + CN^- \rightarrow SO_3^{2-} + SCN^-$$

$$S_4O_6^{2-} + CN^- + H_2O \rightarrow S_2O_3^{2-} + SO_4^{2-} + 2H^+ + SCN^-$$

As shown in Table 13.1 and in the preceding reactions, there are multiple reduced sulfur species that can arise when sulfide minerals are oxidized. There are then multiple pathways for cyanide to be irreversibly consumed as the sulfur species further react during cyanidation to form thiocyanate. The end result can be an appreciable consumption of both oxygen and cyanide. By examining the types of sulfide minerals present in an ore, some general insights can be gained regarding their potential impact on cyanidation. However, only by conducting metallurgical test work can these impacts be quantified for a particular ore being considered for cyanidation. Pyrrhotite and marcasite are known to be particularly problematic in regard to consuming both oxygen and cyanide.

Cyanide Detoxification

The sulfur dioxide (SO_2) and air process is widely used for detoxification of cyanide solutions and slurries. The process was developed by INCO Ltd. in the 1980s and is often called the *INCO SO_2-air process*. It has been proven effective at many mine sites worldwide. The process utilizes SO_2 and air in the presence of a soluble copper catalyst to oxidize cyanide to the less toxic cyanate (OCN^-).

$$SO_2 + H_2O \rightarrow SO_3^{2-} + 2H^+$$
(dissolution of SO_2 to form SO_3^{2-})

$$SO_3^{2-} + O_2 + CN^- \rightarrow OCN^- + SO_4^{2-}$$
(cyanide oxidation)

$$SO_2 + O_2 + H_2O + CN^- \rightarrow OCN^- + SO_4^{2-} + 2H^+$$
(net reaction)

The most widespread application is with slurry tailings, but the process is also effective for the treatment of clear solutions for the oxidation of free and weak acid dissociable (WAD) cyanide. Iron-cyanides are removed through precipitation as insoluble copper-iron-cyanide complexes. Alternative (and equally effective) reagents to SO_2 include sodium metabisulfite ($Na_2S_2O_5$) and ammonium bisulfite (NH_4HSO_3). As shown in the following reactions, the active compound in each of these reagents is sulfite (SO_3^{2-}).

$$Na_2S_2O_5 + H_2O \rightarrow 2Na^+ + 2SO_3^{2-} + 2H^+$$

$$NH_4HSO_3 \rightarrow NH_4^+ + SO_3^{2-} + H^+$$

It may seem counterintuitive to utilize the reducing agent SO_2 to oxidize cyanide, and indeed the process was discovered by accident. But in the presence of oxygen and a copper catalyst, it was found that SO_2 reacts to form an intermediate that is effective at oxidizing cyanide.

Other processes that are used for cyanide detoxification include hydrogen peroxide, Caro's acid, alkaline chlorination, iron-cyanide precipitation, natural attenuation, and aerobic biological. Occasionally, the acidification-volatilization-reabsorption (AVR) stripping-absorption process is used to recover and recycle cyanide.

Source: Mudder et al. 2001; Botz et al. 2016

Cyanide Analysis 14

CLASSIFICATIONS OF CYANIDE

Cyanide complexes may be broadly classified using the terms *free cyanide*, *WAD cyanide*, and total cyanide. WAD cyanide is the widely used acronym for weak acid dissociable or weak and dissociable cyanide.[*] Free cyanide, by definition, consists only of the cyanide anion CN^- and aqueous hydrogen cyanide $HCN_{(aq)}$. Chapter 2 provides a discussion of the forms and chemistry of free cyanide. The distinction between WAD cyanide and total cyanide relates to the stabilities of the various metal-cyanide complexes involved. By stability, it is meant the tendency for a metal-cyanide complex to dissociate at low pH to yield free cyanide in solution. For example, the zinc-cyanide complexes readily dissociate at an acidic pH, while the iron-cyanide complexes are quite stable and dissociate only under conditions of strong acidity, elevated temperature, or exposure to ultraviolet (UV) light.

The relationship between free, WAD, and total cyanide is illustrated in Figure 14.1. As shown, WAD cyanide includes free cyanide plus those metal-cyanide complexes with weak to moderate levels of stability. Note that WAD cyanide is not restricted to just a specific group of metal-cyanide complexes but includes several metal-cyanide complexes plus free cyanide. Total cyanide includes free cyanide plus the metal-cyanide complexes that dissociate under more aggressive chemical conditions.[†] It is true, therefore, that the concentration of WAD cyanide is always equal to or greater than the concentration of free cyanide. Similarly, the concentration of total cyanide is always equal to or greater than the concentration of WAD cyanide.

WAD AND TOTAL CYANIDE

Table 14.1 shows the percent recoveries of cyanide from various metal-cyanide complexes when using the flow injection ligand exchange (FILE) WAD cyanide analytical method.[‡] As shown,

[*] *Available cyanide* is used in the analytical method OIA-1677 (EPA 2010) and is functionally the same as WAD cyanide. Occasionally, total cyanide is referred to as *strong acid dissociable (SAD) cyanide*.

[†] From an analytical perspective, cyanide is known as a *method-defined analyte*. That is, the measured concentration of cyanide depends on the specific analytical method being utilized. Different analytical methods may yield varying recoveries (measurements) of the different metal-cyanide complexes.

[‡] The term *recovery* means the ability of the analytical method to measure the concentration of cyanide associated with each of the metal-cyanide complexes.

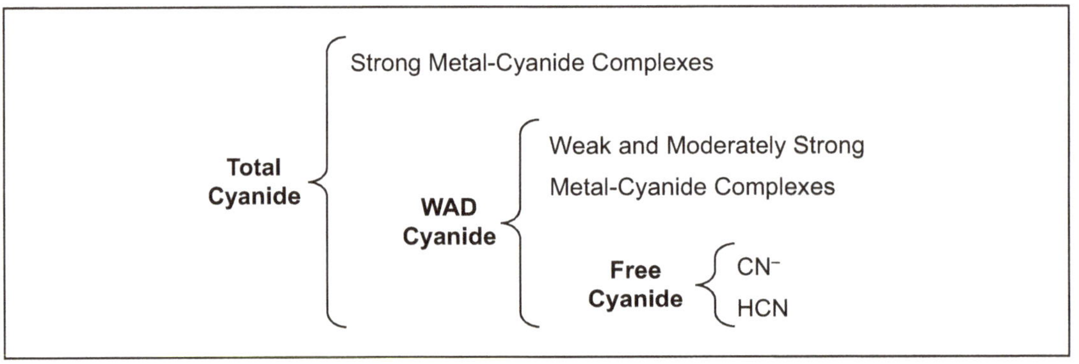

FIGURE 14.1 **Relationship between free, weak acid dissociable cyanide (WAD), and total cyanide**

TABLE 14.1 WAD cyanide recoveries from metal-cyanide complexes*

Complex	WAD Cyanide Recovery, %
$Zn(CN)_4^{2-}$	97.4–98.5
$Cd(CN)_4^{2-}$	100
$Cu(CN)_4^{3-}$	99–100.9
$Ag(CN)_4^{3-}$	100–101.8
$Ni(CN)_4^{2-}$	103–104.3
$Hg(CN)_4^{2-}$	99–100
$Hg(CN)_2^{0}$	98–103.4
$Fe(CN)_6^{4-}$	0
$Fe(CN)_6^{3-}$	0
$Au(CN)_2^{2-}$	0–1.3
$Co(CN)_6^{3-}$	2–2.9†

* WAD = weak acid dissociable or weak and dissociable cyanide.
† Commercial product contains some free cyanide.
Source: EPA 1998

there is essentially 100% recovery of cyanide from the complexes of Zn, Cd, Cu, Ag, Ni, and Hg. The recoveries of cyanide from the complexes of Fe, Au, and Co are essentially zero.

Table 14.2 shows the percent recoveries of cyanide from various metal-cyanide complexes when using total cyanide analytical methods. As shown, there is essentially 100% recovery of cyanide from the complexes of Zn, Cd, Cu, Ag, Ni, Hg, and Fe. The recovery from Au is partial (~50%), while the recovery from Co is essentially zero without exposure to UV light and ~15% with exposure to UV light. The low recovery of cyanide from cobalt is thought to be a result of the slow kinetics for dissociation of cobalt-cyanide. In most leach solutions, the concentrations of Au and Co are relatively low, and the cyanide associated with these metals usually accounts for only a minor amount of the total cyanide in solution. Special ion chromatography techniques can be used to measure the concentrations of the Au-CN and Co-CN complexes.

There are a number of widely accepted analytical methods utilized by laboratories for measuring the concentrations of WAD cyanide and total cyanide. See, for example, the *Cyanide*

TABLE 14.2 Total cyanide recoveries from metal-cyanide complexes

Complex	Total Cyanide Recovery, %
$Zn(CN)_4^{2-}$	97.2–104.4
$Cd(CN)_4^{2-}$	100.9–104.3
$Cu(CN)_4^{3-}$	97.7–100.3
$Ag(CN)_2^{-}$	97.8–104.3
$Ni(CN)_4^{2-}$	97.1–104.2
$Hg(CN)_4^{2-}$	95.8–99.6
$Hg(CN)_2^{0}$	96.1–102.5
$Fe(CN)_6^{4-}$	99.5–101
$Fe(CN)_6^{3-}$	83.5–104
$Au(CN)_2^{2-}$	42.8–58.4
$Co(CN)_6^{3-}$	0–2.5 (no UV*); 13.8–14.6 (with UV)

* UV = ultraviolet light

Data from Solujic et al. 1999

Analysis Guide (OI Analytical, n.d.) and "An Overview and Comparison of Methods for Cyanide Analysis" (OI Analytical 2009). ASTM also has a guidance document that describes interferences and method selection for compliance with the International Cyanide Management Code (ASTM D7728).

When implementing a cyanide analytical method at a leaching operation, seek the advice of instrument suppliers or consultants who can guide you through the proper selection of instruments and techniques. There are many options available with regard to cost, accuracy, detection limits, ability to quantify different cyanide species, susceptibility to interferences, automation, and complexity. Several online process instruments are also available to measure titratable cyanide and WAD cyanide in leach solution.

SILVER NITRATE TITRATION

Some commentary is in order regarding the silver nitrate ($AgNO_3$) titration method that is widely used to measure the concentration of cyanide in leach solution. In fact, this method is probably used at nearly all cyanidation operations worldwide. It has the significant advantages of being simple, quick, inexpensive, and easy to understand. In many cases, the method is suitably accurate and can be used to control the dosing of sodium cyanide (NaCN) into leach circuits. The titration method is often referenced as a way of measuring the concentration of free cyanide, but this is not always the case.

There are some aspects of the silver nitrate titration method that should be understood by operators. The method is based on the following reaction of silver with cyanide (nitrate does not take part in the reaction).

$$Ag^+ + 2CN^- \rightleftharpoons Ag(CN)_2^-$$

In this reaction, two moles of cyanide always react with one mole of silver. As silver nitrate is added during the titration, the preceding reaction continues until all the available cyanide has reacted with silver to form the complex $Ag(CN)_2^-$. At this point any further addition of silver nitrate results in an excess of Ag^+ in solution. It is at exactly this point that the titration is complete and has reached the end point. The end point is usually detected visually by a color change using an indicator such as potassium iodide (KI) or more commonly 5-(4-dimethyl-aminobenzylidene) rhodanine (simply referred to as *rhodanine* or *rhodanine-B*). Occasionally, a potentiometric probe is used for detection of the end point. The amount of silver nitrate added until the end point is reached is then used to back-calculate the amount of cyanide that was titrated, according to the ratio of two moles cyanide per one mole of silver. For solutions containing only free cyanide, the chemistry is straightforward and the results are quite accurate.

However, the situation can significantly change if metals are present in solution along with cyanide. The cyanide complexes of gold, iron, cobalt, nickel, and mercury are relatively stable and do not react to any notable extent with silver nitrate. Cadmium is normally present at low levels and the amount of cyanide associated with cadmium is usually of little consequence relative to process control. And of course, the silver-cyanide complex does not react any further during a silver nitrate titration. This leaves copper and zinc.

Zinc

Provided the pH during the titration is sufficiently elevated (and it should be by adding 1 M sodium hydroxide [NaOH] to give a pH >12), all of the cyanide associated with zinc will be titrated. It is obvious then that the titration, in the case of zinc being present, over-reports the concentration of free cyanide because the zinc-cyanide complexes are not free cyanide. Usually, this is not a significant hinderance to using the titration method with leach solutions containing zinc. But operators need to recognize free cyanide is not being measured and that the results are qualitative (or empirical) rather than quantitative. That is, trends in the results and operational experience can be used as a guide to control dosing of NaCN into a leach circuit. But the free cyanide being reported by the titration method is not truly a measurement of free cyanide, rather it is a measurement of titratable cyanide. Elevated zinc concentrations in leach solutions are most often associated with Merrill–Crowe circuits, though occasionally elevated zinc concentrations arise from the leaching of zinc-containing minerals.

Copper

The issue is more complex with copper, and increasingly widespread. The copper-cyanide complexes are $Cu(CN)_2^-$, $Cu(CN)_3^{2-}$, and $Cu(CN)_4^{3-}$ and all of these complexes are present (along with some level of free cyanide) in leach solutions containing copper. During the silver nitrate titration, it is thought that silver fully reacts with free cyanide and partially reacts with $Cu(CN)_3^{2-}$ and $Cu(CN)_4^{3-}$ but not with $Cu(CN)_2^-$. In the author's experience, the extent of the reaction of silver nitrate with the copper-cyanide complexes depends on the concentrations of cyanide and copper, the molar ratio of cyanide to copper (CN:Cu), the amount of dilution water added to the sample before titration, the amount of color indicator added, and the rate at which silver nitrate is added. Color changes are also subjective and the visually observed end points often differ from person to person. This frequently results in variable, or even erratic, measurements of cyanide when copper is present. If elevated thiosulfate levels are also present, this can further add to the variability (thiosulfate falsely increases the measured level of titratable cyanide). In short, with leach solutions containing elevated levels of copper, the silver nitrate titration method often does not provide a quantitative measurement of cyanide. But as with zinc, trends in the titration results and operational experience can still sometimes be used as a guide to control dosing of NaCN into a leach circuit.

The author has observed on several occasions leach solutions thought to have sufficient free cyanide levels for leaching, when in fact the free cyanide level was very low (essentially zero). Leaching of gold and silver in these cases was inhibited by the cyanide-starved leach conditions. This was caused by silver nitrate titrations falsely suggesting free cyanide was present, whereas the titration was actually reporting some of the cyanide bound to copper. This is an important issue when dealing with elevated levels of copper in leach solution, particularly when activated carbon is being used for gold recovery.

This issue with copper can be remedied somewhat by using an automatic titrator with potentiometric end point detection (Breuer et al. 2011). An automatic titrator yields more repeatable readings and possibly higher-quality results closer to the true free cyanide concentration, but the method likely will not provide a true measurement of free cyanide. A modified silver nitrate titration method (modified potentiometric [MP]-WAD) is available that provides good measurements of WAD cyanide in process solutions at low cost (Botz et al. 2013). The technique behind the MP-WAD titration can also be used with online automated silver nitrate titrators to give measurements of WAD cyanide rather than measurements of titratable cyanide.

If an accurate estimation of the true free cyanide concentration is needed, the suggested approach is first to measure the concentrations of WAD cyanide (or total cyanide) and the metals in solution, most importantly copper and zinc. Then the detailed speciation of the individual cyanide complexes can be calculated using an equilibrium calculator. Visual MINTEQ is free software and can be used for this purpose. But others are available, such as Geochemist's Workbench, STABCAL, and HSC Chemistry.

An example is given in Table 14.3 for a solution containing 747 mg/L WAD cyanide and 500 mg/L copper at pH 10.5 and 25°C (measured values). The titratable cyanide concentration

Natural Sources of Cyanide

Cyanide is naturally found in a number of foods and plants, including almonds, millet sprouts, lima beans, soy, spinach, bamboo shoots, and cassava roots (which are a major source of food in tropical countries). However, the edible parts of plants eaten in the United States, including tapioca, which is made from cassava roots, contain relatively low amounts of cyanide. Certain bacteria, fungi, and algae can also produce cyanide. Vitamin B_{12}, a natural chemical containing cyanide, is beneficial to your body because it prevents anemia (iron-poor blood). The cyanide binds in vitamin B_{12} so that it does not serve as a source of cyanide exposure and cannot harm you. Smoking is probably one of the major sources of cyanide exposure for people who do not work in cyanide-related industries.

Source: ATSDR 2006

TABLE 14.3 Calculated speciation of cyanide complexes

Parameter	Value
pH	10.5
Temperature (°C)	25
WAD cyanide (mg/L as CN)	747
Copper (mg/L as Cu)	500
Molar ratio CN:Cu	3.65:1
$AgNO_3$ Titration	**Concentration, mg/L as CN**
Potentiometric end point	110
Rhodanine end point	257
Calculated Speciation	**Concentration, mg/L as CN**
CN^-	94
$HCN_{(aq)}$	4
$Cu(CN)_2^-$	<1
$Cu(CN)_3^{2-}$	510
$Cu(CN)_4^{3-}$	138
Sum	747

for this solution was reported to be 110 mg/L using a potentiometric end point and 257 mg/L using a rhodanine end point (Breuer et al. 2011). The true free cyanide concentration calculated by the author for this solution is 98 mg/L (94 mg/L CN^- plus 4 mg/L $HCN_{(aq)}$). The potentiometric end point for this particular solution provided a reasonably accurate estimate of the free cyanide concentration, but the rhodanine endpoint yielded a concentration about 2.6 times higher than the true free cyanide concentration. These results do not suggest the potentiometric end point always provides an accurate estimate of the free cyanide concentration. Regardless of the end point technique used with the conventional silver nitrate titration, when copper is present in solution the results should be viewed as empirical and not necessarily reflective of the true free cyanide concentration.

There is a useful technique for estimating the titratable cyanide concentration using detailed speciation results, similar to those presented in Table 14.3. First, it is known that the titratable cyanide concentration includes free cyanide and all the cyanide associated with zinc. Second, cyanide associated with gold, silver, iron, cobalt, nickel, cadmium, and mercury can be ignored. Third, it can then be assumed (as an approximation) that cyanide associated with copper will be titrated down to a final CN:Cu molar ratio of 2.4:1 (Breuer et al. 2011). For example, with the 500 mg/L of copper shown in Table 14.3, the following is calculated using the assumed final CN:Cu molar ratio of 2.4:1 when the titration is complete:

$$\frac{500\ \cancel{\text{mg Cu}}}{\text{L}} \times \frac{\cancel{\text{mmol Cu}}}{63.546\ \cancel{\text{mg Cu}}} \times \frac{2.4\ \cancel{\text{mmol CN}}}{\cancel{\text{mmol Cu}}} \times \frac{26.0174\ \text{mg CN}}{\cancel{\text{mmol CN}}} = \frac{491\ \text{mg CN}}{\text{L}}$$

Thus, at the end of the titration using rhodanine color indicator, there remains an estimated 491 mg/L of cyanide bound to copper (i.e., copper-cyanide that was not measured by the titration). Per Table 14.3, the amount of cyanide bound to copper before the titration is 510 + 138 = 648 mg/L. The amount of copper-cyanide titrated by silver nitrate is therefore estimated to be 648 – 491 = 157 mg/L. Cyanide in this sample that was titrated by silver nitrate is then calculated to be 94 mg/L (CN^-) + 4 mg/L ($HCN_{(aq)}$) + 157 mg/L ($Cu(CN)_3^{2-}$ and $Cu(CN)_4^{3-}$) = 255 mg/L. This is in good agreement with the value of 257 mg/L shown in Table 14.3.

As another example, consider the pregnant solution chemistry shown in Table 12.1. The free cyanide and zinc-cyanide concentrations sum to 264 mg/L as CN for this solution, while there is 501 mg/L of cyanide bound to copper before the titration (see Table 12.2). The copper concentration of 400 mg/L corresponds to 393 mg/L of cyanide remaining bound to copper at the end of the titration, following the same calculation sequence as just shown. Therefore, titration of this solution using rhodanine color indicator is estimated to yield a concentration of 264 + (501 – 393) = 372 mg/L as titratable cyanide.

Properties of Sodium Cyanide

15

Cyanide utilized at gold and silver leaching operations is most often purchased as sodium cyanide (NaCN), either as a bulk solid or in concentrated solution form. Solid sodium cyanide is nonvolatile; white in color; and available as granules, flakes, or more commonly, briquettes. The solid may be shipped in 1 t (metric ton) steel bins or in 1 t wooden boxes with two-layer internal plastic bagging. The solid can also be shipped in larger steel isotanks with 20–24 t capacity. After arrival at a mine site, water is recirculated through the isotanks to dissolve the sodium cyanide, which is then transferred as a solution into a fixed storage tank.

The purity of industrial grade solid sodium cyanide is 97%–99% NaCN, with the principal impurities being sodium hydroxide (NaOH; 0.1%–0.5%), sodium carbonate, sodium formate, and water. The concentrated solution is clear to slightly opaque and usually contains 28%–32% NaCN along with 0.5%–2% NaOH. The solution is delivered in bulk in about 18–20 t quantities.

Some basic properties of solid sodium cyanide are given in Table 15.1.

TABLE 15.1 Basic properties of solid sodium cyanide

Property	Value
Molecular weight	49.007
Specific gravity	1.6
Bulk density	880–960 kg/m^3

Data from DuPont 2005

Sodium cyanide is readily soluble in water up to a maximum concentration of about 38 wt % at 25°C. Solubilities at other temperatures are shown in Table 15.2.

TABLE 15.2 Solubility of sodium cyanide in water

Temperature, °C	Solubility, wt %
0	29
5	30.3
10	31.9
15	33.6
20	35.9
25	38.1
30	41
35	44.9

Data from DuPont 2005

A solubility diagram for the NaCN-H_2O system is shown in Figure 15.1. At low temperatures and concentrations of <25% NaCN, there is a region where ice is formed in the solution. At higher concentrations (>25% NaCN), the solid NaCN·2H_2O can form in the solution. Solid-free solutions of NaCN exist within the shaded region shown in Figure 15.1.

The freezing temperatures of sodium cyanide solutions as a function of concentration are shown in Figure 15.2. The data in Figure 15.2 are the same as in Figure 15.1 but with the axes reversed. There is a eutectic point that occurs

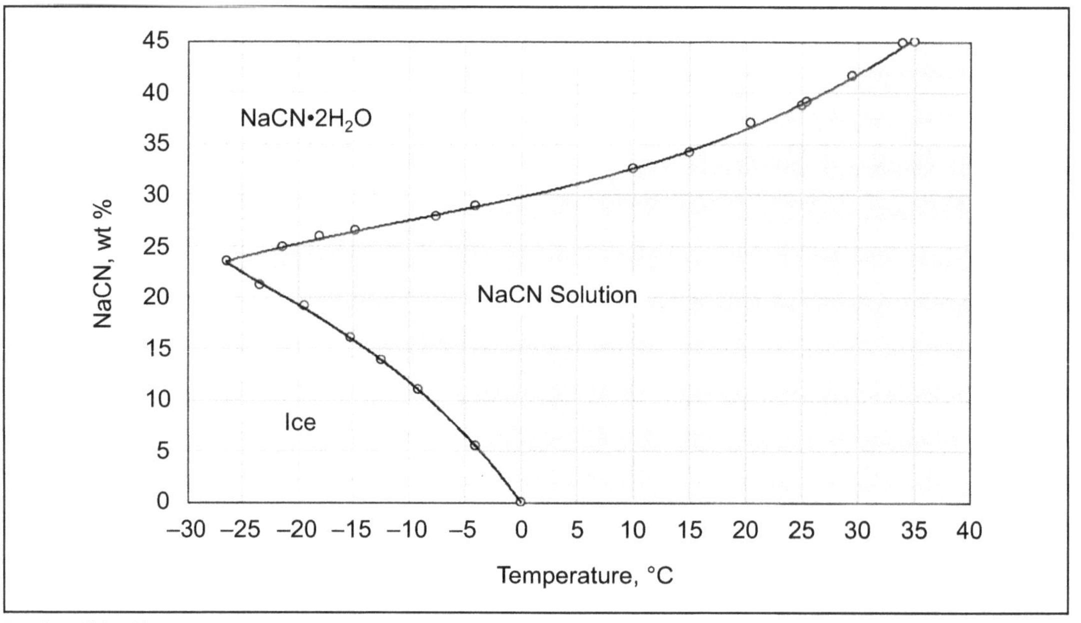

Data from Linke 1965

FIGURE 15.1 Solubility diagram for the NaCN-H_2O system

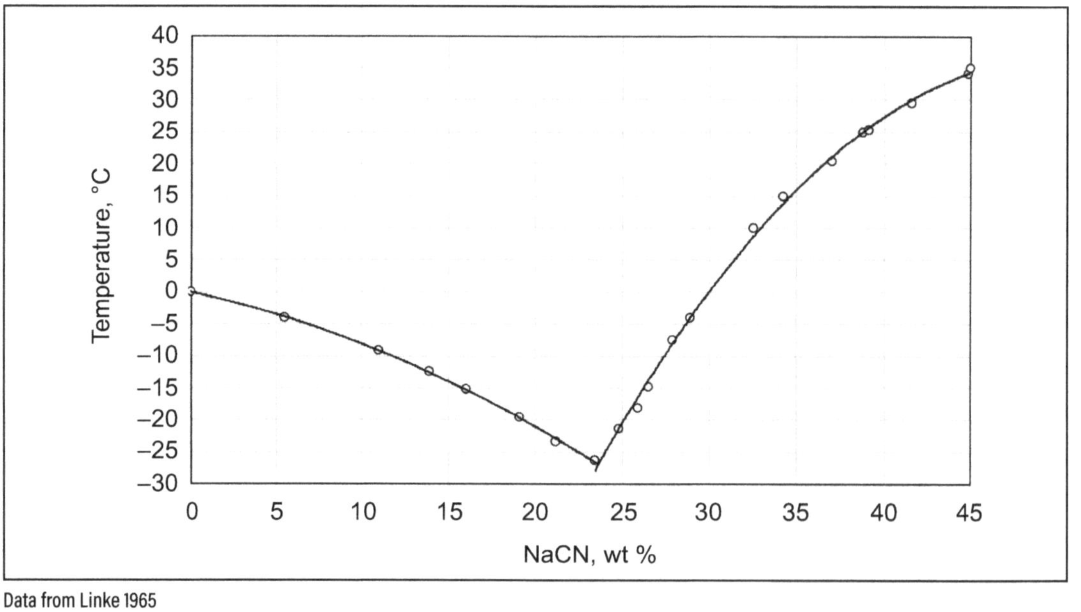

Data from Linke 1965

FIGURE 15.2 Freezing temperatures of sodium cyanide solutions

at a concentration of 23.5% NaCN where the minimum freezing temperature is about −26°C. The freezing temperatures of the solutions are slightly lowered by the addition of 0.5%–2.0% NaOH.

The densities of sodium cyanide solutions at a range of concentrations and temperatures are shown in Table 15.3.

TABLE 15.3 Density of sodium cyanide solutions

NaCN, wt %	Density, g/cm³	
	20°C*	25°C†
5	1.04	1.02
10	1.07	1.05
15	1.1	1.07
20	1.12	1.1
25	1.15	1.12
30	1.17	1.15
35	1.19	1.18

* Orica, n.d.
† Dupont 2005

The viscosities of sodium cyanide solutions at a range of concentrations and temperatures are shown in Tables 15.4 and 15.5. The data in Tables 15.4 and 15.5 come from different sources and are not in good agreement, but this could be caused by different impurities present in the solutions considered. In general, sodium cyanide solutions are readily pumped, as the viscosity is moderate even at higher concentrations (*Ullmann's Encyclopedia of Industrial Chemistry* 1987).

When dissolving sodium cyanide into fresh water, a small amount of sodium hydroxide should first be added to ensure a sufficiently high pH. A high pH reduces the tendency for hydrogen cyanide gas to evolve. Data available from DuPont suggests solutions of 10%–30% NaCN should include a minimum of 0.3%–0.5% sodium hydroxide to provide a protective pH level of 12.5–13.0. An increased amount of sodium hydroxide is not harmful to the solution and further reduces the tendency for hydrogen cyanide gas to evolve. The equilibrium levels of

TABLE 15.4 Viscosity of sodium cyanide solutions

NaCN, wt %	Viscosity, cP			
	0°C	10°C	20°C	30°C
5	1.9	1.4	1.1	0.9
10	2.1	1.6	1.2	1
15	2.4	1.8	1.4	1.1
20	2.9	2.1	1.6	1.3
25	3.6	2.7	2.1	1.6
30	4.6	3.4	2.6	2.1

Data from Orica, n.d.

TABLE 15.5 Viscosity of 30% sodium cyanide solution

Temperature, °C	Viscosity, cP
21.5	10.3
30	7.6
50	2
80	0.9

Data from CSBP 2016

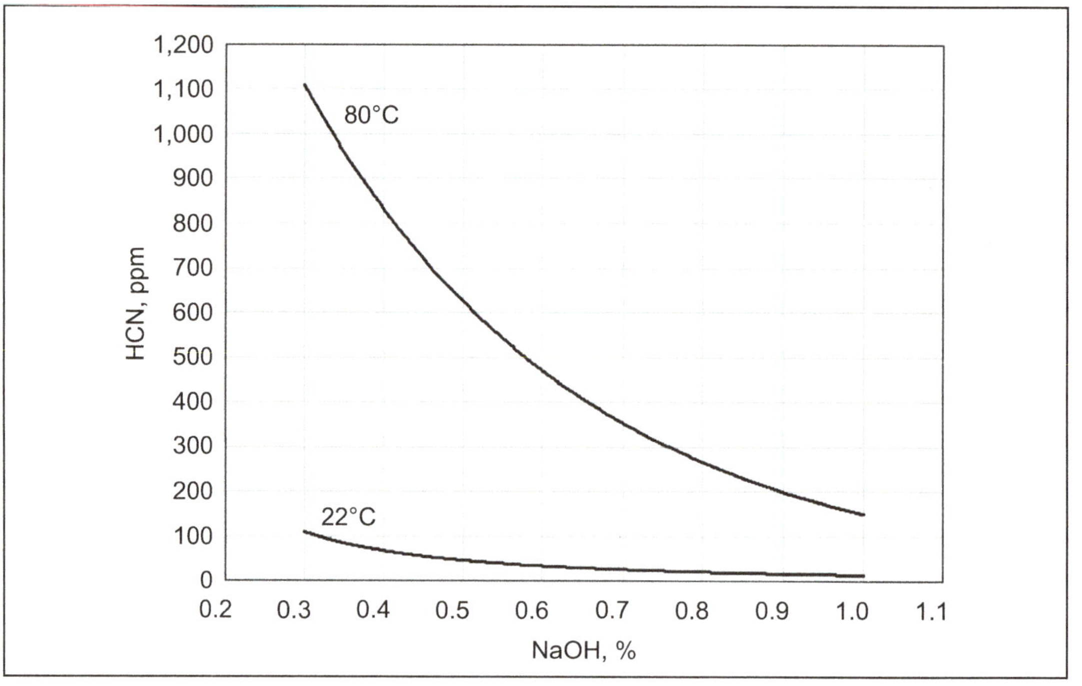

Data from Degussa, n.d.

FIGURE 15.3 HCN concentrations above 30% NaCN solution

hydrogen cyanide gas above a 30% solution of sodium cyanide are shown in Figure 15.3 for temperatures of 22°C and 80°C.

Solutions of sodium cyanide are relatively stable and only decompose slowly when stored at ambient temperature. Information published by DuPont (2005) indicates at temperatures of 20°–30°C the decomposition rate is roughly 0.03%–0.10% per day. The decomposition rate increases rapidly when the temperature rises above approximately 30°C. Solutions with higher levels of sodium hydroxide tend to decompose more slowly, while higher temperatures promote more rapid decomposition. Degussa (n.d.) reports a 27%–30% NaCN solution stored at 30°C loses 1.4% of the NaCN content over a period of 84 days (i.e., about 0.02% per day). Similarly, Orica (n.d.) reports a decomposition rate of about 0.03%–0.06% per day at 25°C.

When solid NaCN is dissolved into fresh water, the reaction is slightly endothermic, meaning that the solution cools slightly. It is similar to sodium chloride (NaCl) in this respect. For practical purposes, the change in temperature can be ignored when preparing solutions from dry NaCN. This is opposite to the behavior of sodium hydroxide (NaOH), which is well known to have a high heat of dilution and yields solutions that are quite warm when prepared from the dry solid. Shown in Table 15.6 is a comparison of the heats of dilution for NaCN, NaCl, and NaOH.

TABLE 15.6 Heat of solution in water at infinite dilution

Compound	ΔH_{SOL}, kJ/mol
NaCN	+1.21 (endothermic)
NaCl	+3.88 (endothermic)
NaOH	−44.51 (exothermic)

Source: Yaws 2012–2014

Specialty Cyanide Compounds

Cyanide compounds can be purchased for specialty chemistry purposes in a wide variety of solid forms, such as those listed below. These compounds are occasionally used at gold and silver leaching operations or in metallurgical testing laboratories for calibration of analytical instruments or for conducting specialized test work.

Barium cyanide	$Ba(CN)_2$
Calcium cyanide	$Ca(CN)_2$
Copper (I) cyanide	$CuCN$
Gold (I) cyanide	$AuCN$
Iron (III) ferrocyanide (Prussian blue)	$Fe_4(Fe(CN)_6)_3$
Mercury (II) cyanide	$Hg(CN)_2$
Nickel (II) cyanide tetrahydrate	$Ni(CN)_2{\cdot}4H_2O$
Potassium cyanide	KCN
Potassium dicyanoaurate (I)	$KAu(CN)_2$
Potassium hexacyanocobaltate (III)	$K_3Co(CN)_6$
Potassium hexacyanoferrate (III)	$K_3Fe(CN)_6$
Potassium hexacyanoferrate(II) trihydrate	$K_4Fe(CN)_6{\cdot}3H_2O$
Potassium tetracyanonickelate (II) hydrate	$K_2Ni(CN)_4{\cdot}xH_2O$
Sodium cyanide	$NaCN$
Silver cyanide	$AgCN$
Zinc cyanide	$Zn(CN)_2$

Cyanide Safety

16

Experience at many leaching operations has proven cyanide can be safely handled in large quantities. However, safe handling requires that personnel be provided with proper training, procedures, and protective equipment. Reputable suppliers of cyanide can assist with these efforts and should be consulted in this regard. Suppliers can also provide information about suitable equipment for cyanide storage and pumping systems. The International Cyanide Management Institute has a code of practice that should be followed by manufacturers, transporters, and users of cyanide (ICMI 2021).

Cyanide is well known as a toxic and fast-acting acute poison. Dangerous exposures can occur through inhalation, ingestion, and absorption through the skin. A high level of diligence is required when handling cyanide in solid or solution form. The toxicity of hydrogen cyanide gas to humans is shown in Table 16.1. The International Cyanide Management Code references HCN worker exposure limits of 10 ppm maximum and 4.7 ppm over 8 hours (ICMI 2021). Operations often have fixed and portable gas monitors to alarm at these levels, or at lower levels, as may be appropriate according to local regulations or company policy. The odor threshold for hydrogen cyanide is reported as 2–5 ppm, although about one-third of persons are not sensitive to the odor.

Birds, wildlife, and livestock may be exposed to cyanide that is present in open process ponds, such as barren/pregnant solution ponds and tailings storage facilities. The International Cyanide Management Code identifies 50 mg/L weak acid dissociable (WAD) cyanide as the limit in open waters to protect these organisms (ICMI 2021).

TABLE 16.1 Toxicity of hydrogen cyanide in air to humans

Concentration, ppm	Response
0-17	No observed effect
20-40	Minimal symptoms after several hours
45-54	Tolerated for 30-60 minutes
90	Fatal after 60 minutes
135	Fatal after 30 minutes
181	Fatal after 10 minutes
270	Immediately fatal

Source: NIOSH 1976

Aquatic life is very sensitive to cyanide, and the levels must be low in waters discharged to the environment. The International Cyanide Management Code identifies 0.022 mg/L free cyanide as the limit in receiving waters downstream of a mixing zone to protect aquatic life (ICMI 2021). Local limits may be more stringent, and local limits may also reference WAD cyanide or total cyanide rather than free cyanide.

All persons working around cyanide must be familiar with its hazards and observe established safety practices. Among the basic safety practices that should be observed are the following:

- Always follow company policies for personal protective equipment (PPE). Distinct levels of PPE may be required for the different tasks performed at cyanide leaching operations.
- Do not breathe sodium cyanide (NaCN) dust or solution mist. Where there is a risk of inhalation, an approved respirator must be worn.
- Avoid contact of cyanide, in both dry and solution form, with the skin, eyes, open wounds, or abrasions. Butyl rubber and neoprene gloves are effective and have low permeabilities, but neoprene is more rugged. Eye goggles are recommended whenever there is a risk of splashing.
- Prevent the contact of cyanide with acids and never store acids near cyanide. It is good practice to have fixed and portable hydrogen cyanide gas monitors in areas where cyanide is handled in solution form.
- Use a broom and shovel to clean spills of dry NaCN; avoid the use of wash water to the extent practical.
- Do not eat, drink, or smoke in areas where cyanide is present, and do not store food or drink in these areas.
- Store solid NaCN in a dry, ventilated, and secure area with restricted access. Do not store under water sprinklers, if allowed by code and local regulations. If sprinklers are required, ensure that adequate secondary containment is installed.
- Develop and test emergency plans and equipment for responding to incidents involving cyanide.

Information from DuPont (2005), Australian Government (2008), Logsdon et al. (1999), and Safe Work Australia (2013) is available at no cost, and these references provide valuable information about cyanide safety and management.

Supplemental Data

The following tables (A.1 through A.6) provide supplemental data that readers may find useful. Also included here is a periodic table of the elements (Figure A.1)

TABLE A.1 Composition of dry atmospheric air (sea level)

Gas Species	Molecular Weight	Volume, %
N_2	28.0134	78.084
O_2	31.9988	20.9476
Ar	39.948	0.934
CO_2	44.0095	0.0314
Ne	20.1797	0.001818
He	4.002602	0.000524
Kr	83.798	0.000114
Xe	131.293	0.0000087
CH_4	16.04246	0.0002
H_2	2.01588	0.00005
Average	28.9644	—

Source: NOAA 1976

TABLE A.2 Atmospheric air pressure versus elevation

Elevation, m	Pressure, bar
0	1.01325
500	0.95461
1,000	0.89876
1,500	0.84559
2,000	0.79501
2,500	0.74691
3,000	0.70121
3,500	0.6578
4,000	0.6166
4,500	0.57752
5,000	0.54048
5,500	0.50539
6,000	0.47217

Source: NOAA 1976

TABLE A.3 Solubility of oxygen in water (sea level)

Temperature, °C	Oxygen Solubility, mg/L	
	Total Dissolved Solids, 0 mg/L	Total Dissolved Solids, 10,000 mg/L
0	14.6	13.6
5	12.8	11.9
10	11.3	10.6
15	10.1	9.5
20	9.1	8.6
25	8.3	7.8
30	7.6	7.2
35	6.9	6.6
40	6.4	6.1

Source: Benson and Krause 1984

TABLE A.4 Commonly used conversion units*

From	To	Multiply By
ft^3	gal (US)	7.480519481
gal (US)	L	3.785411784
inch	mm	25.4
kg	lb	2.204622622
lb	g	453.59237
lb	oz troy	14.58333333
m	ft	3.280839895
m^3	gal (US)	264.1720524
oz troy	g	31.1034768
oz/ton	g/t	34.28571429
ton	lb	2,000
t	ton	1.102311311

*** ton = shoret ton; t = metric tona**

TABLE A.5 Standard sieve sizes

U.S. Mesh (ASTM E11-22)	Size, mm	Tyler Mesh	Size, mm
4	4.75	4	4.699
6	3.35	6	3.327
8	2.36	8	2.362
10	2	10	1.651
12	1.7	12	1.397
14	1.4	14	1.168
16	1.18	16	0.991
20	0.85	20	0.833
30	0.6	28	0.589
35	0.5	35	0.417
40	0.425	48	0.298
50	0.3	65	0.212
60	0.25	100	0.147
70	0.212	150	0.104
80	0.18	200	0.074
100	0.15	270	0.052
140	0.106	325	0.044
170	0.09	400	0.037
200	0.075	—	—
270	0.053	—	—
325	0.045	—	—
400	0.038	—	—

TABLE A.6 Stability constants for common cyanide species (25°C)

Species	log β_1	log β_2	log β_3	log β_4	log β_5	log β_6
HCN	9.22	—	—	—	—	—
Au(I)	18	38.92	—	—	—	—
Ag(I)	9.7	20.48	21.7	—	—	—
Fe(II)	—	—	—	—	—	47
Fe(III)	—	—	—	—	—	52
Co(III)	—	—	—	—	62	70.04
Ni(II)	—	—	—	30.2	—	—
Cu(I)	—	23.94	29.24	30.74	—	—
Zn(II)	5.34	11.07	16.05	19.62	—	—
Cd(II)	6.01	11.12	15.65	17.92	—	—
Hg(II)	17	32.75	36.31	38.97	—	—

1 **H** Hydrogen 1.00794																	2 **He** Helium 4.002602
3 **Li** Lithium 6.941	4 **Be** Beryllium 9.0121831											5 **B** Boron 10.811	6 **C** Carbon 12.0107	7 **N** Nitrogen 14.0067	8 **O** Oxygen 15.9994	9 **F** Fluorine 18.99840316	10 **Ne** Neon 20.1797
11 **Na** Sodium 22.98976928	12 **Mg** Magnesium 24.305											13 **Al** Aluminum 26.9815385	14 **Si** Silicon 28.0855	15 **P** Phosphorus 30.973762	16 **S** Sulfur 32.065	17 **Cl** Chlorine 35.453	18 **Ar** Argon 39.948
19 **K** Potassium 39.0983	20 **Ca** Calcium 40.078	21 **Sc** Scandium 44.955908	22 **Ti** Titanium 47.867	23 **V** Vanadium 50.9415	24 **Cr** Chromium 51.9961	25 **Mn** Manganese 54.938044	26 **Fe** Iron 55.845	27 **Co** Cobalt 58.933194	28 **Ni** Nickel 58.6934	29 **Cu** Copper 63.546	30 **Zn** Zinc 65.38	31 **Ga** Gallium 69.723	32 **Ge** Germanium 72.63	33 **As** Arsenic 74.9216	34 **Se** Selenium 78.971	35 **Br** Bromine 79.904	36 **Kr** Krypton 83.798
37 **Rb** Rubidium 85.4678	38 **Sr** Strontium 87.62	39 **Y** Yttrium 88.90584	40 **Zr** Zirconium 91.224	41 **Nb** Niobium 92.90637	42 **Mo** Molybdenum 95.95	43 **Tc** Technetium 98	44 **Ru** Ruthenium 101.07	45 **Rh** Rhodium 102.9055	46 **Pd** Palladium 106.42	47 **Ag** Silver 107.8682	48 **Cd** Cadmium 112.414	49 **In** Indium 114.818	50 **Sn** Tin 118.71	51 **Sb** Antimony 121.76	52 **Te** Tellurium 127.6	53 **I** Iodine 126.90447	54 **Xe** Xenon 131.293
55 **Cs** Cesium 132.905452	56 **Ba** Barium 137.327		72 **Hf** Hafnium 178.49	73 **Ta** Tantalum 180.94788	74 **W** Tungsten 183.84	75 **Re** Rhenium 186.207	76 **Os** Osmium 190.23	77 **Ir** Iridium 192.217	78 **Pt** Platinum 195.084	79 **Au** Gold 196.966569	80 **Hg** Mercury 200.59	81 **Tl** Thallium 204.3833	82 **Pb** Lead 207.2	83 **Bi** Bismuth 208.9804	84 **Po** Polonium 209	85 **At** Astatine 210	86 **Rn** Radon 222
87 **Fr** Francium 223	88 **Ra** Radium 226																
			57 **La** Lanthanum 138.90547	58 **Ce** Cerium 140.116	59 **Pr** Praseodymium 140.90766	60 **Nd** Neodymium 144.242	61 **Pm** Promethium 145	62 **Sm** Samarium 150.36	63 **Eu** Europium 151.964	64 **Gd** Gadolinium 157.25	65 **Tb** Terbium 158.92535	66 **Dy** Dysprosium 162.5	67 **Ho** Holmium 164.93033	68 **Er** Erbium 167.259	69 **Tm** Thulium 168.93422	70 **Yb** Ytterbium 173.054	71 **Lu** Lutetium 174.9668
			89 **Ac** Actinium 227	90 **Th** Thorium 232.0377	91 **Pa** Protactinium 231.03588	92 **U** Uranium 238.02891	93 **Np** Neptunium 237	94 **Pu** Plutonium 244	95 **Am** Americium 243	96 **Cm** Curium 247	97 **Bk** Berkelium 247	98 **Cf** Californium 251	99 **Es** Einsteinium 252	100 **Fm** Fermium 257	101 **Md** Mendelevium 258	102 **No** Nobelium 259	103 **Lr** Lawrencium 260

FIGURE A.1 Periodic table of the elements

Further Reading

The following are offered as excellent resources for further reading in regard to gold and silver cyanidation chemistry and associated metallurgical processes.

Dzombak, D.A., Ghosh, R.S., and Wong-Chong, G.M. 2006. *Cyanide in Water and Soil: Chemistry, Risk, and Management.* Boca Raton, FL: CRC Press.

Flynn, C.M., and McGill, S.L. 1995. *Cyanide Chemistry: Precious Metals Processing and Waste Treatment.* Washington, DC: U.S. Bureau of Mines.

Hedley, N., and Tabachnick, H. 1958. *Chemistry of Cyanidation.* Mineral Dressing Notes No. 23. Wayne, NJ: American Cyanamid Company.

Lorösch, J. 2001. *Process and Environmental Chemistry of Cyanidation.* Hong Kong: Degussa AG.

Marsden, J.O., and House, C.I. 2006. *The Chemistry of Gold Extraction.* Littleton, CO: SME.

Mudder, T.I., Botz, M.M., and Smith, A. 2001. *Chemistry and Treatment of Cyanidation Wastes,* 2nd ed. London: Mining Journal Books.

Stanley, G.G., ed. 1987. *The Extractive Metallurgy of Gold in South Africa,* vols. 1–2. Johannesburg: Chamber of Mines of South Africa.

Yannopoulos, J.C. 1991. *The Extractive Metallurgy of Gold.* New York: Van Nostrand Reinhold.

References

Adams, M.D. 2001. A methodology for determining the deportment of cyanide losses in gold plants. *Minerals Engineering* 14(4):383–391.

Adams, M.D., and Kyle, J.H. 2000. Precipitation of cyanide as $Cu_2Fe(CN)_6$ compounds from cyanidation and detoxification circuits. In *MINPREX 2000.* Melbourne, Victoria: Australasian Institute of Mining and Metallurgy.

American Cyanamid. 1953. *The Chemistry of the Ferrocyanides.* Wayne, NJ: American Cyanamid Company.

Aqueous Solutions. 2011–2023. The Geochemist's Workbench [software application]. www.gwb.com. Accessed December 2023.

ASTM D7728. 2011. *Standard Guide for Selection of ASTM Analytical Methods for Implementation of International Cyanide Management Code Guidance.* West Conshohocken, PA: ASTM International.

ASTM E11-22. 2022. *Standard Specification for Woven Wire Test Sieve Cloth and Test Sieves.* West Conshohocken, PA: ASTM International.

ATSDR (Agency for Toxic Substances and Disease Registry). 2006. *Toxicological Profile for Cyanide.* Washington, DC: U.S. Department of Health and Human Services, Public Health Service.

Australian Government.2008. *Cyanide Management*. www.industry.gov.au/sites/default/files/2019–04/lpsdp-cyanide-management-handbook-english.pdf. Accessed December 2023.

Baes, C.M., and Mesmer, R.E. 1976. *The Hydrolysis of Cations*. Malabar, FL: Krieger.

Bailey, P.R. 1987. Application of activated carbon to gold recovery. In *The Extractive Metallurgy of Gold in South Africa* 1. Edited by G.G. Stanley. Johannesburg: Chamber of Mines of South Africa. pp. 379–614.

Bard, A.J., Parsons, R., and Jordan, J., eds. 1985. *Standard Potentials in Aqueous Solution*. Boca Raton, FL: CRC Press.

Benson, B.B., and Krause, D. 1984. The concentration and isotopic fractionation of oxygen dissolved in freshwater and seawater in equilibrium with the atmosphere. *Limnology and Oceanography* 29(3):620–632.

Bethke, C.M. 2008. *Geochemical and Biogeochemical Reaction Modeling*, 2nd ed. Cambridge, UK: Cambridge University Press.

Botz, M.M., Kaczmarek, A., and Orser, S. 2011. Managing copper in leach solution at the Çöpler gold mine: Laboratory testing and process design. *Minerals & Metallurgical Processing* 28(3):133–138.

Botz, M.M., Milosavljevic, E.B., and Ward, I. 2013. Measurement of weak and dissociable (WAD) cyanide with a modified potentiometric titration. *Minerals & Metallurgical Processing* 30(4):197–204.

Botz, M.M., Mudder, T.I., and Akcil, A.U. 2016. Cyanide treatment: Physical, chemical, and biological processes. In *Advances in Gold Ore Processing*. Edited by M.D. Adams. New York: Elsevier. pp. 672–702.

Botz, M.M., Phillips, W., Polglase, T., et al. 2001. Technologies for the regeneration of cyanide from thiocyanate. *Minerals & Metallurgical Processing* 18(3):126–132.

Breuer, P.L., Sutcliffe, C.A., and Meakin, R.L. 2011. Cyanide measurement by silver nitrate titration: Comparison of rhodanine and potentiometric end-points. *Hydrometallurgy* 106:135–140.

Broderius, S.J. 1973. Determination of molecular hydrocyanic acid in water and studies of the chemistry and toxicity of metal-cyanide complexes. PhD dissertation, Oregon State University, Corvalis, OR.

Butler, J.N. 1998. *Ionic Equilibrium: Solubility and pH Calculations*. Hoboken, NJ: Wiley.

Charette, M.A., and Smith, W.H.F. 2010. The volume of Earth's ocean. *Oceanography* 23(2):112–114.

Christensen, J.J., Izatt, R.M., and Eatough, D.J. 1965. Thermodynamics of metal cyanide coordination. Log K, ΔH°, and ΔS° values for the Hg2+-CN system. *Inorganic Chemistry* 4(9):1278–1280.

Clennell, J.E. 1915. *The Cyanide Handbook*, 2nd ed. New York: McGraw-Hill.

CSBP. 2016. Safety data sheet, sodium cyanide 30% solution. Kwinana, Western Australia: CSBP Limited (Australian Gold Reagents).

Czehura, S.J. 2006. Butte: A world class ore deposit. *Mining Engineering* 58(9):14–19, 25.

Dean, J.A., ed. 1992. *Lange's Handbook of Chemistry*, 14th ed. New York: McGraw-Hill.

Degussa. n.d. Sodium cyanide, potassium cyanide properties, reactions, handling. Hong Kong: Degussa AG.

Deschênes, G., Rajala, J., Pratt, A.R., et al. 2011. Advances in the cyanidation of silver. *Minerals & Metallurgical Processing* 28(1):37–43.

DuPont. 2005. *Sodium Cyanide Properties, Uses, Storage, and Handling.* Wilmington, DE: DuPont.

Dzombak, D.A., Ghosh, R.S., and Wong-Chong, G.M. 2006. *Cyanide in Water and Soil: Chemistry, Risk, and Management.* Boca Raton, FL: CRC Press.

Edwards, T.J., Maurer, G., Newman, J., et al. 1978. Vapor-liquid equilibria in multicomponent aqueous solutions of volatile weak electrolytes. *AIChE Journal* 24(6):966–976.

EPA (U.S. Environmental Protection Agency). 1998. 40 CFR Part 136: Guidelines establishing test procedures for the analysis of pollutants; available cyanide. *Federal Register* 63(129):36810–36824.

EPA (U.S. Environmental Protection Agency). 2006. MINTEQA2 equilibrium speciation model, version 4.3 [software]. Washington, DC: EPA, Center for Exposure Monitoring.

EPA (U.S. Environmental Protection Agency). 2010. *Method OIA-1677-09: Available Cyanide by Ligand Exchange and Flow Injection Analysis (FIA).* Washington, DC: EPA.

Flynn, C.M., and McGill, S.L. 1995. *Cyanide Chemistry: Precious Metals Processing and Waste Treatment.* Washington, DC: U.S. Bureau of Mines.

Ford-Smith, M.H. 1964. *The Chemistry of Complex Cyanides: A Literature Survey.* London: Her Majesty's Stationery Office.

Fuerstenau, M.C., and Han, K.N., eds. 2003. *Principles of Mineral Processing.* Littleton, CO: SME.

Gamsjäger, H., Bugajski, J., Gajda, T., et al. 2005. Chemical thermodynamics of nickel. In *Chemical Thermodynamics,* vol. 6. Edited by Nuclear Energy Agency. Amsterdam, The Netherlands: Elsevier Science.

Goh, S.W., Buckley, A.N., Lamb, R.N., Rosenberg, R.A., and Moran, D. 2006. The oxidation states of copper and iron in mineral sulfides, and the oxides formed on initial exposure of chalcopyrite and bornite to air. *Geochimica et Cosmochimica* (70): 2210–2228.

Gray, T. 2009. *The Elements: A Visual Exploration of Every Known Atom in the Universe.* New York: Black Dog & Leventhal.

Greenwood, N.N., and Earnshaw, A. 1984. *Chemistry of the Elements.* Oxford, UK: Pergamon Press.

Habashi, F. 1967. *Kinetics and Mechanisms of Gold and Silver Dissolution in Cyanide Solution.* Bulletin 59. Butte: Montana Bureau of Mines and Geology.

Hallock, W., and Herbert, W.T. 1906. *Outlines of the Evolution of Weights and Measures and the Metric System.* New York: Macmillan.

Hampton, A.P. 2002. Zinc cementation—The Merrill-Crowe process. In *Mineral Process Plant Design, Practice, and Control* 2. Edited by A.L. Mular, D.N. Halbe, and D.J. Barratt. Littleton, CO: SME. pp. 1663–1679.

Hedley, N., and Tabachnick, H. 1958. *Chemistry of Cyanidation.* Mineral Dressing Notes No. 23. Wayne, NJ: American Cyanamid Company.

Hsu, Y.J., and Tran, T. 1996. Selective removal of gold from copper-gold cyanide liquors by cementation using zinc. *Minerals Engineering* 9(1):1–13.

ICMI (International Cyanide Management Institute). 2021. *International Cyanide Management Code, Mining Operations Verification Protocol, Standards of Practice 4.4 and 4.5.* Washington, DC: ICMI.

Izatt, R.M., Christensen, J.J., Hansen, et al. 1965. Thermodynamics of metal cyanide coordination. I: Log βi°, ΔHi°, and ΔSi° values for the zinc(II)-cyanide system at 25°. *Inorganic Chemistry* 4(5):718–721.

Izatt, R.M., Johnston, H.D., Watt, G.D., et al. 1967. Thermodynamics of metal cyanide coordination. VI: Copper(I)- and silver(I)-cyanide systems. *Inorganic Chemistry* 6(1):132–135.

Izatt, R.M., Johnston, H.D., Eatough, D.J., et al. 1971. Thermodynamics of metal cyanide coordination. Part IX: Log K, ΔH°, and ΔS° values for the Ni2+-, Zn2+-, Cd2+-, and Hg2+- CN- systems at 10, 25, and 40°C. *Thermochimica Acta* 2:77–85.

Knovel. 2008. pKsp of low solubility inorganic solutes. In *Critical Tables*, 2nd ed. https://app.knovel.com/resources/kpKCTE000X/toc. Accessed December 2023.

Kraft, A. 2008. On the discovery and history of Prussian blue. *Bulletin for the History of Chemistry* 33(2):61–67.

Krauskopf, K.B. 1967. *Introduction to Geochemistry*. New York: McGraw-Hill.

Leaver, E.S., and Woolf, J.A. 1929. Effect of *Copper and Zinc in Cyanidation with Sulfide-Acid Precipitation*. Technical Publication No. 250, October. New York: American Institute of Mining and Metallurgical Engineers.

Lide, D.R., ed. 2005. *CRC Handbook of Chemistry and Physics*, 86th ed. Boca Raton, FL: CRC Press.

Lingle, J. 2000. *Complexation and Solubility Reactions Controlling Cyanide Environmental Behavior*. [Electric Power Research Institute] EPRI Document 1000588, November. Palo Alto, CA: EPRI.

Linke, W.F. 1965. *Solubilities Inorganic and Metal-Organic Compounds*, 4th ed., vol. 2, Washington, DC: American Chemical Society. p. 909.

Logsdon, M.J., Hagelstein, K., and Mudder, T.I. 1999. *The Management of Cyanide in Gold Extraction*. Melbourne, Victoria, Australia: International Council on Metals and the Environment.

Lorösch, J. 2001. *Process and Environmental Chemistry of Cyanidation*. Hong Kong: Degussa AG.

Lower, G.W., and Booth, R.B. 1965. Recovery of copper by cyanidation. *Mining Engineering* (November): 56–60.

Lu, J., Dreisinger, D.B., and Cooper, W.C. 2002. Thermodynamics of the aqueous copper-cyanide system. *Hydrometallurgy* 66:23–26.

Luna, R.M., and Lapidus, G.T. 2000. Cyanidation kinetics of silver sulfide. *Hydrometallurgy* 56:171–188.

Luthy, R.G., and Bruce, S.G. 1979. *Kinetics of Reactions of Cyanide and Reduced Sulfur Species to Form Thiocyanate*. Washington, DC: U.S. Department of Energy.

MacFarren, H.W. 1912. *Text Book of Cyanide Practice*. New York: McGraw-Hill.

Marks, L. 2000. *Between Silk and Cyanide: A Codemaker's War 1941–1945*. New York: Free Press.

Marsden, J.O., and House, C.I. 2006. *The Chemistry of Gold Extraction*. Littleton, CO: SME.

Maxwell, G.R. 2004. *Synthetic Nitrogen Products: A Practical Guide to the Products and Processes*. New York: Plenum.

Metso Outotec. n.d. HSC Chemistry Software, version 8.1.4 [software application]. www.metso.com/portfolio/hsc-chemistry/. Accessed December 2023.

Montana Tech (Montana Technological University). n.d. STABCAL [Stability Calculations for Aqueous Systems] software application. Butte, MT: Department of Metallurgical and Materials Engineering.

Mudder, T.I., Botz, M.M., and Smith, A. 2001. *Chemistry and Treatment of Cyanidation Wastes*, 2nd ed. London: Mining Journal Books.

NBS (National Bureau of Standards). 1982. *Tables of Chemical Thermodynamic Properties*. Washington, DC: NBS.

Nicol, M.J., Fleming, C.A., and Paul, R.L. 1987. The chemistry of the extraction of gold. In *The Extractive Metallurgy of Gold in South Africa* 2. Edited by G.G. Stanley. Johannesburg: Chamber of Mines of South Africa. pp. 831–905.

NIOSH (National Institute for Occupational Safety & Health). 1976. *Criteria for a Recommended Standard: Occupational Exposure to Hydrogen Cyanide and Cyanide Salts*. Publication No. 77–108. Pittsburgh, PA: NIOSH.

NIST (National Institute of Standards and Technology). 2004. *NIST Standard Reference Database 46: Critically Selected Stability Constants of Metal Complexes (NIST46)*. Gaithersburg, MD: NIST.

NOAA (National Oceanic and Atmospheric Administration). 1976. U.S. standard atmosphere. Technical memorandum NOAA–S/T 76–1562. Washington, DC: U.S. Government Printing Office.

OI Analytical. n.d. *Cyanide Analysis Guide*. Publication 29680212. College Station, TX: OI Analytical. www.xylem.com/siteassets/brand/oi-analytical/resources/brochures/oi-analytical-cyanide-analysis-guide-brochure.pdf

OI Analytical. 2009. An overview and comparison of methods for cyanide analysis. Presented at the 2009 Pittsburgh Conference on Analytical Chemistry and Applied Spectroscopy, Chicago, IL, March 8–13.

Orica. n.d. *Orica Cyanide Technical Manual*. East Melbourne, Victoria, Australia: Orica Mining Chemicals.

Osseo-Asare, K., Xue, T., and Ciminelli, S.T. 1984. Solution chemistry of cyanide leaching systems. In *Precious Metals: Mining, Extraction, and Processing*. Edited by V. Kudryk, D.A. Corrigan, and W.W. Liang. Littleton, CO: SME. pp. 173–197.

Parsons, P., and Dixon, G. 2013. *The Periodic Table: A Visual Guide to the Elements*. London: Quercus.

Pourbaix, M. 1974. *Atlas of Electrochemical Equilibria in Aqueous Solution*. 2nd ed. Houston, TX: National Association of Corrosion Engineers.

Rees, K.L., and van Deventer, J.S.J. 2000. Preg-robbing phenomena in the cyanidation of sulphide gold ores. *Hydrometallurgy* 58:61–80.

Safe Work Australia. 2013. *Guide for Preventing and Responding to Cyanide Poisoning in the Workplace*. April. Canberra: Safe Work Australia.

Sharpe, A.G. 1976. *The Chemistry of Cyano Complexes of the Transition Metals*. Cambridge, MA: Academic Press.

Sibrell, P.L., Atkinson, G.B., and Walters, L.A. 1994. *Cyanide Leaching Chemistry of Platinum-Group Metals*. Report of Investigation 9507. Washington, DC: U.S. Bureau of Mines.

Sillén, L.G. 1964. *Stability Constants Of Metal-Ion Complexes. Section I: Inorganic Ligands*. Special Publication No. 17. London: The Chemical Society.

Smith, R.M., and Martell, A.E. 1976. Inorganic complexes. In *Critical Stability Constants*, vol. 4. New York: Plenum Press.

Solujic, L., Milosavljevic, E.B., and Straka, M.R. 1999. Total cyanide determination by a segmented flow injection—On-line UV digestion—Amperometric method. *Analyst* 124:1255–1260.

Special Metals Corporation. n.d. Home page. www.specialmetals.com/. Accessed December 2023.

Suzuki, I. 1999. Oxidation of inorganic sulfur compounds: chemical and enzymatic reactions. *Canadian Journal of Microbiology* 45:97–105.

TTB (U.S. Alcohol and Tobacco Tax and Trade Bureau). 2023. Wine labeling: Declaration of sulfites. www.ttb.gov/labeling-wine/wine-labeling-declaration-of-sulfites#:~:text=Sulfites%20are%20preservatives%20widely%20used,back%20label%20of%20the%20product. Accessed December 2023.

Ullmann's Encyclopedia of Industrial Chemistry. 1987. s.v. "Cyano compounds, inorganic." Hoboken, NJ: Wiley. pp. 159–190.

van der Merwe, W., and Breuer, P. 2013. Cyanide analysis for complex cyanide solutions. *CIM Journal* 4(2):120–128.

Wasserlauf, M., and Dutrizac, J.E. 1982. *The Chemistry, Generation and Treatment of Thiosalts in Milling Effluents—A Non-Critical Summary of CANMET Investigations 1976–1982.* CANMET Report 82–4E. Ottawa, ON: Canada Centre for Mineral and Energy Technology. p. 982.

WHO (World Health Organization). 2013. *Model List of Essential Medicines*, 18th ed. Geneva, Switzerland: WHO.

Wilmot, J.C. 1997. The chemistry of cyanide in the presence of reduced sulfur sources. PhD dissertation, University of Nebraska, Lincoln, NE.

Yannopoulos, J.C. 1991. *The Extractive Metallurgy of Gold.* New York: Van Nostrand Reinhold.

Yaws, C.L., ed. 2012–2014. *Yaws' Critical Property Data for Chemical Engineers and Chemists.* https://app.knovel.com/kn/resources/kpYCPDCECD/toc. Accessed December 2023.

Index

Note: *f.* indicates figure; *t.* indicates table

acanthite, 19–20
acetone cyanohydrin, 11
adiponitrile, 11
agitated tank leaching operations, 2
ammonia odor, carbon elution and, 36
available cyanide, 73

barium cyanide, 85
batteries, 57
Between Silk and Cyanide (Marks), 45
Butte, MT, 26

cadmium, 55–57
  Eh-pH diagrams, 56*f.*–57*f.*
  in leach solutions, 3*t.*, 4, 55
  pH-distribution diagram, 56*f.*
  SART process, 40
  stability constants, 55*t.*
  total cyanide recovery, 75*t.*
  uses for, 57
  WAD cyanide recovery, 74*t.*
calcium cyanide, 3, 12, 85
carbon elution, 36
chalcopyrite, 41, 43
cinnabar, 59
cobalt, 29–31
  Eh-pH diagrams, 30*f.*–31*f.*
  in leach solutions, 3*t.*, 4
  pH-distribution diagram, 31*f.*
  stability constants, 29*t.*
  total cyanide recovery, 75*t.*
  uses for, 57
  WAD cyanide recovery, 74*t.*
cobaltite, 29
copper, 37–44
  copper-sulfide mineral composition, 44*t.*
  Eh-pH diagrams, 39*f.*, 43*f.*–44*f.*
  in leach solutions, 3*t.*, 4, 41*f.*–42*f.*
  mineral solubilities, 38*t.*
  as a nuisance metal, 38
  pH-distribution diagram, 40*f.*
  prices, 38n
  from richest hill on earth, 26
  SART process, 40
  silver nitrate titration, 76–79, 78*t.*
  stability constants, 37*t.*
  total cyanide recovery, 75*t.*
  WAD cyanide recovery, 74*t.*
copper cyanide, 3, 5, 85
Crowe, T.B., 49
Crowe towers, 49
cyanate, 4–5, 5*f.*
cyanide
  chemical structure, 2*f.*
  defined, 2
  natural sources of, 77
Cyanide Mountain, 8
cyanide-starved solutions, 39

Dearborn River, 8
detoxification, 2, 5, 71
dyes, 27

Eh-pH diagrams, 5–8, *7f.*
Elbow Creek, 8
Elsner's equation, 15
Entombment of Christ (Werff), 27
espionage, 45

free cyanide, 9–10, *10f.*, *10t.*, 73, *74f.*, 88

gallium, 65
gold, 15–17
  carbon elution, 36
  Eh-pH diagram, *16f.*
  gold-tellurides, 15
  leach reaction, 15
  in leach solutions, *3t.*, 4
  leaching recoveries, 1–2
  preference for adsorption, 17, *17f.*
  prices, 38n
  properties of, 12
  from richest hill on earth, 26
  stability constants, *15t.*
  total cyanide recovery, *75t.*
  troy system, 16
  WAD cyanide recovery, *74t.*
  world's largest deposit, 17
gold cyanide, 2, 85
golden superlatives, 17

heap leach operations, 2
hematite, 25–26
Henry's law constants, 11, *11t.*
Homestake mine, 49
hydrogen cyanide, 2, 9–11, *10f.*, *87t.*
hydrogen sulfide, 68

INCO SO_2-air process, 71
industrial cyanide, 2–3
International Cyanide Management Institute, 87–88
invisible ink, 57
iron, 23–28
  Eh-pH diagrams, 24–28, *25f.*, *27f.–28f.*
  leach reactions, 23–24, *24t.*
  in leach solutions, *3t.*, 4
  stability constants, *23t.*
  total cyanide recovery, *75t.*
  WAD cyanide recovery, *74t.*
iron ferrocyanide, 85

Le Chatelier's principle, 9
leach solutions, 2
leaching, 2
  pregnant leach solution example, 63–65, *63t.–64t.*
lead, 26
lead nitrate, 19–20
ligands, 2, 5
linnaeite, 29

manganese, 26
marcasite, 70
Marks, Leo, 45
mercury, 59–61
  Eh-pH diagrams, *60f.–61f.*
  in leach solutions, *3t.*, 4, 59
  pH-distribution diagram, *60f.*
  retorting, 65
  SART process, 40
  stability constants, *59t.*
  total cyanide recovery, *75t.*
  WAD cyanide recovery, *74t.*
mercury cyanide, 85
Merrill, C.W., 49
Merrill-Crowe process, 49
metal-cyanide complexes, 3–8, *3t.*, 69, 73–75, *74f.*, *74t.–75t.*
methionine, 11
method-defined analyte, 73n
molecular weights for basic cyanide compounds, *12t.*

molybdenum, 26
Montana Mining Company, 49

nature, cyanide in, 77
nickel, 33–35
  Eh-pH diagrams, *34f.*–*35f.*
  in leach solutions, *3t.*, 4
  pH-distribution diagram, *34f.*
  stability constants, *33t.*
  total cyanide recovery, *75t.*
  uses for, 57
  WAD cyanide recovery, *74t.*
nickel cyanide tetrahydrate, 85

oceans, gold in, 17
oxidation reactions, 6

pentlandite, 33
pH
  Eh-pH diagrams, 5–8, *7f.*
  free cyanide, 9–10, *10f.*
pigments, 27
plastics production, 11
poison antidotes, 27
polysulfides, 69, 70
polythionates, 69, 70
potassium cyanide, 3, 12, 85
potassium dicyanoaurate, 85
potassium hexacyanocobaltate, 85
potassium hexacyanoferrate, 3, 85
potassium hexacyanoferrate trihydrate, 85
potassium tetracyanonickelate hydrate, 85
Pourbaix, Marcel, 5n
Pourbaix diagrams, 5n
pregnant leach solution example, 63–65, *63t.*–*64t.*
Prussian blue, 2, 26–27, *27f.*–*28f.*, 85
pyrite, 41
pyrrhotite, 70

radioactive poisoning, 27
recovery, defined, 73n
recycling, cyanide, 2
redox, 6
reduced sulfur species, 67, *67t.*
reduction reactions, 6
richest hill on Earth, 26
Rio de la Plata, 12

safety considerations, 87–88, *87t.*
SART process, 40
selenocyanate, 4
Sharpe, A.G., 4
silver, 19–21
  carbon elution, 36
  Eh-pH diagrams, *20f.*–*21f.*
  leach reaction, 19–20
  in leach solutions, *3t.*, 4
  leaching recoveries, 1–2
  pH-distribution diagram, *21f.*
  properties of, 12
  from richest hill on earth, 26
  SART process, 40
  stability constants, *19t.*
  total cyanide recovery, *75t.*
  troy system, 16
  WAD cyanide recovery, *74t.*
silver cyanide, 85
silver nitrate titration, 13, 64–65, 75–79
simple cyanides, 12
smaltite, 29
sodium cyanide, 2–3, 11, 12, 81–84, *81t.*, *82f.*, *83t.*–*84t.*, *84f.*, 85
sodium hydroxide, 2
specialty compounds, 85
spies, 45
stability constants, 5
strong acid dissociable (SAD) cyanide, 73n
sulfate, 67
sulfide minerals, 4–5, 67–70

sulfidization-acidification-recycling-thickening (SART) process, 40
sulfite, 69
sulfur, 67–70, 67*t.*, 68*f.*
sulfur dioxide, 71

tarnish, 12
tetrathionate, 69
thiocyanate, 4–5, 5*f.*, 68
thiosulfate, 19, 69, 70
total cyanide, 73–75, 74*f.*
trithionate, 69
troy system, 16
Turnbull's blue, 2, 26–27, 27*f.*–28*f.*

unit conversions, 12–13

volatilization, hydrogen cyanide, 10–11

weak acid dissociable (WAD), 13
weak acid dissociable (WAD) cyanide, 73–75, 74*f.*, 74*t.*–75*t.*, 87
Werff, Pieter van der, 27
wine, sulfites in, 69

zinc, 47–53
  cementation, 47, 49, 50, 52*t.*, 53, 59
  Eh-pH diagrams, 48*f.*, 52*f.*
  in leach solutions, 3*t.*, 4
  mineral solubilities, 48*t.*, 51*f.*
  pH-distribution diagram, 49*f.*
  from richest hill on earth, 26
  SART process, 40
  silver nitrate titration, 76
  stability constants, 47*t.*
  total cyanide recovery, 75*t.*
  WAD cyanide recovery, 74*t.*
zinc cyanide, 85